Physical Chemistry

Physical Chemistry

Diana Lucas

Larsen & Keller
www.larsen-keller.com

Physical Chemistry
Diana Lucas
ISBN: 978-1-64172-633-7 (Hardback)

Larsen & Keller
Published by Larsen and Keller Education,
5 Penn Plaza,
19th Floor,
New York, NY 10001, USA

Cataloging-in-Publication Data

Physical chemistry / Diana Lucas.
p. cm.
Includes bibliographical references and index.
ISBN 978-1-64172-633-7
1. Chemistry, Physical and theoretical. 2. Chemistry. I. Lucas, Diana.
QD453.3 .P49 2021
541--dc23

For more information regarding Larsen and Keller Education and its products, please visit the publisher's website www.larsen-keller.com

Table of Contents

Preface

The study of macroscopic, subatomic, atomic and particulate phenomena related to chemical systems is referred to as physical chemistry. It is primarily concerned with the application of principles, concepts and practices of physics to chemical systems. Some of these are force, time, motion, energy, thermodynamics, quantum chemistry, statistical mechanics, chemical equilibrium and analytical dynamics. Major sub-domains of physical chemistry are quantum chemistry, spectroscopy, chemical thermodynamics and chemical kinetics. It is also involved in the study of surface science, electrochemistry of cell membranes, the effect of reaction kinetics on the rate of a reaction and intermolecular forces that act upon the physical properties of materials. This book provides comprehensive insights into the field of physical chemistry. Some of the diverse topics covered herein address the varied branches that fall under this category. Those in search of information to further their knowledge will be greatly assisted by this book.

A foreword of all Chapters of the book is provided below:

Chapter 1 - The branch of chemistry which applies the theories and techniques from physics to study chemical systems is known as physical chemistry. Some of the important concepts within physical chemistry are Henry's law, Raoult's law and Hess's law. This is an introductory chapter which will introduce briefly all these significant topics of physical chemistry; **Chapter 2 -** The hypothetical gases whose molecules do no interact with each other and occupy negligible space are known as ideal gases. The real gases have interactions between their molecules and thus do not behave like ideal gases. Some of the laws which govern gases are Boyle's law and Charles law. This chapter has been carefully written to provide an easy understanding of the varied facets of ideal and real gases; **Chapter 3 -** The branch of chemistry which deals with the phenomena which occur at the surfaces of substances is known as surface chemistry. Some of these phenomena are adsorption and molecular diffusion. The topics elaborated in this chapter will help in gaining a better perspective about the topics which are studied under surface chemistry; **Chapter 4 -** The sub-field of chemistry which deals with the application of quantum mechanics in experiments and physical models of chemical systems is known as quantum chemistry. Some of concepts which are studied within it are electronic structure and chemical dynamics. This chapter discusses in detail these theories and concepts related to quantum chemistry; **Chapter 5 -** The state where the properties of a system do not exhibit any observable change due to the rate of forward reaction equaling the rate of backward reaction is known as a chemical equilibrium. Some of the principles which are used to study chemical equilibriums are Le-Chatelier's principle and Nernst distribution law. The chapter closely examines these key concepts of equilibrium chemistry to provide an extensive understanding of the subject.

I would like to thank the entire editorial team who made sincere efforts for this book and my family who supported me in my efforts of working on this book. I take this opportunity to thank all those who have been a guiding force throughout my life.

Diana Lucas

1

Introduction to Physical Chemistry

The branch of chemistry which applies the theories and techniques from physics to study chemical systems is known as physical chemistry. Some of the important concepts within physical chemistry are Henry's law, Raoult's law and Hess's law. This is an introductory chapter which will introduce briefly all these significant topics of physical chemistry.

Physical chemistry is the branch of chemistry concerned with the interpretation of the phenomena of chemistry in terms of the underlying principles of physics. It lies at the interface of chemistry and physics, inasmuch as it draws on the principles of physics (especially quantum mechanics) to account for the phenomena of chemistry. It is also an essential component of the interpretation of the techniques of investigation and their findings, particularly because these techniques are becoming ever more sophisticated and because their full potential can be realized only by strong theoretical backing. Physical chemistry also has an essential role to play in the understanding of the complex processes and molecules characteristic of biological systems and modern materials.

It also involves the study of the properties of substances at different scales, from the macroscopic scale which includes particles that are visible to the naked eye, to the subatomic scale involving extremely small subatomic particles such as electrons.

Physical chemistry differs from other branches of chemistry because it employs the concepts and principles of physics to understand chemical systems and reactions. The different scales that this branch of chemistry deals with are described below.

The Macroscopic Scale

The macroscopic scale involves the substances that are large enough to be visible to the human eye (without the help of optical instruments to provide a magnified view). Some important quantities that are dealt with on the macroscopic scale include:

- Melting and boiling points.
- The coefficient of linear thermal expansion.

- Latent heat of vaporization.
- Enthalpy of fusion.

Considering the example of a water molecule, the macroscopic scale involves its physical state (solid, liquid, or gaseous) but does not deal with the chemistry of the H_2O molecule.

The Microscopic Scale

The microscopic scale deals the with properties of substances that can only be seen by the human eye with the help of optical instruments that magnify the substance, such as microscopes.

The study of the shapes and structures of crystals fall under this scale. The structures of crystals have an impact on the behavior of large sections of the crystals, which are used in bridges and airplanes.

The Atomic Scale

- The properties of matter at the atomic scale vary from element to element.
- However, some elements exhibit similar properties and are grouped together in the periodic table.
- Some examples of the properties of matter studied in physical chemistry at the atomic scale include atomic mass and atomic number.
- The value of the atomic radii of elements can be used to determine the bond length of the bond between two atoms in a molecule.

Subatomic Scale

- The subatomic scale involves the study of particles that are smaller than the size of atoms, generally referred to as subatomic particles.
- At this scale, the dual nature of particles is the reason that subatomic particles are sometimes referred to as waves or energy.
- The study of physical chemistry at a very advanced level involves the study of subatomic particles.
- The study of these particles also have applications in the field of nuclear chemistry.

As the name suggests, physical chemistry uses the techniques, theories, and principles that are associated with physics to explain certain aspects of chemistry. It can be noted that factors such as temperature and pressure have an impact at the macroscopic and the microscopic level but their impact diminishes at the atomic and the subatomic scales.

Branches of Physical Chemistry

The manner in which physics can be applied to explain or solve chemical problems make up the important concepts of physical chemistry.

Thermodynamic and Thermochemistry

Some studies show that heat is the most important factor in chemical reaction. Heat drives some reaction to occur but for some cases it retains reaction. The study about learning how to measure the heat transfer in chemical reaction is a branch of physical science called thermodynamic, which describes the meaning of heat and gives procedures for measuring heat transfer quantitatively. Heat is just another form of energy which known as thermal energy.

At this point, we know that the transfer of heat called thermodynamic which is more a domain of physical. Then, we can say the thermodynamic of chemical aspect is called thermochemistry.

Because chemical reactions are usually studied in constant pressure, heat transfer is reaction are measured at constant pressure. To obtain a number of heat transfer at constant-pressure, a chemical reaction is needed and the amount of the heat transferred symbolized as q_p.

This reaction called reaction enthalpies which as formulation as:

$$q_p = \Delta H = H_f - H_p = H_{product} - H_{reactan} = \Delta H_{reaction}$$

When carbon monoxide burns in oxygen to produce carbon dioxide, the heat is release or given off into the environment, it has a negative sign. When heat is given off by reaction (ΔH is negative), the reaction is said to be exothermic. The contrary reaction of exothermic which heat is taken up showed by ΔH positive are called endothermic. The example of endothermic is the preceding reaction written in the opposite direction or reverse direction of carbon dioxide.

Thermodynamic unable us to predict ΔH of decomposition reaction with complete confidence, even if calorimetric experiment is never actually done of performed for it.

Standard State Enthalpies

The absolute values of the enthalpy of an element behave the same like internal energy means it can not be measured or calculated. Only the changes in enthalpy can be measured. It is necessary to adopt a reference state for enthalpies of substances therefore with this problem chemist define standard states for principles of modern chemistry as follows:

- For solid and liquid, the standard state is the thermodinamycally stable state at a pressure of 1 atm and at a specified temperature.
- For gases, the standard state is the gaseous phase at a pressure of 1 atm, at specified temperature and exhibiting ideal gas behavior.
- For dissolved species, the standard state is a 1-m solution at a pressure of 1 atm, at specified temperature and exhibiting ideal solution behavior.

The most common temperature choosen for specified temperature is 298,15 K (25 C exactly). So, when the temperature standard is not stated therefore the temperature is 298,15 K. Once the standards have been defined, it can be concluded that the chemical element in their standard states at 298,15 K have zero enthalpy.

The delta enthalpy or the change in enthalpy when the conditions are in standard. It means all reactants and products are in their standard state with the specified temperature called standard enthalpy symbolized as ΔH^{o}_{f}. The enthalpy changes for the reaction that produce 1 mol of the compound from its element in their stable states is called standard enthalpy of formation concept ΔH^{o}_{f} of compound.

Bond Enthalpies

To form in a new arrangement of the atom a chemical reaction requires an existing bond to break. Chemists have found that in highly reactive intermediate species there could be more than one bond have been broken and find the amount of energy required to break a particular bond.

Once such important measurable quantity is the enthalpy changes when a bond is broken in the gas phase called the bond enthalpy. This change is invariably positive because atoms bonded together have lower energy than when it's separated. For instance, the reaction happen in bond enthalpy of a C- H in methane need 438kJ mol^{-1}, when measured in standard enthalpy condition.

Thermochemistry Process

A process happens if one or more properties of a system changes. Process could be divided into three categories:

- Isobaric means the process occurs when the pressure of system remaining constant.
- Isothermal means the process occurs when the temperature of system remaining constant.
- Adiabatic means process occurs without any heat exchange.

Electrochemistry

Electrochemistry is the branch of physical chemistry concerned with the inter conversion of chemical and electrical energy through oxidation-reduction reaction. Electrons are transferred between two hall-reaction through external electrical energy. Electrical energy can be used to drive chemical reactions that are not naturally spontaneously occurred.

Electrochemical methods are very widely used in medical diagnostic, where they routinely measure very low concentration of molecules of clinical significance. Other applications of electrochemistry are found in solar energy, conversion, fuel cells and energy storage technology.

Galvanic Cells

Galvanic cells is one of electrochemical cells, electrochemical itself is the location where electrochemical reaction interconvert chemical and electrical energy through coupled reduction oxidation reaction in which half-reaction are separated.

According to Gibbs theory, in galvanic cells free energy made available in spontaneous reduction oxidation reaction and can be converted into electrical energy which can be useful for work. The example of galvanic cell happen on reduction-oxidation reaction of copper metal to form copper ions, sand silver ion have been reduced to silver metal. A wire connects the two metal electrodes, allowing electrons to flow between them and the ionic conductor connecting cells is called a salt bridge.

The other concept must be known is anode and cathode. Anode is the location of electrode when it experience an oxidation and cathode is the location where electrode experience reduction. These terms and concept was stated by Michael Faraday.

Electrolytic Cells

In galvanic cell, all reaction described as spontaneously occurred and the energy change is written negative. This thermodynamic driving force related to an electrostatic driving force. The system composed of charged particles spontaneously seek the of the lowest electrostatic potential energy just like mechanical system spontaneously seek states of the lowest mechanical potential energy. In other word, this electrolytic cell will be activated when an energy applied under reduction-oxidation reaction occur.

Three major parts of electrolytic are the substance that produce an electrically conducting solution or called as electrolyte and the other are electrodes such as anode and cathode. So, we could conclude that Electrolytic cell is working in contrary way of galvanic cell or batteries.

Faraday's Laws

In 1833, Michael Faraday found and demonstrated quantitative relationship between the amount of charge that followed and the quantities of material product or consumed in electrochemical reactions. Therefore the relationship he discovered known as Faraday's Law which state as follows:

- The mass of a given substance that is produce or consume in electrochemical reaction is proportional to the quantity of electric charge passed.
- Equivalent masses of different substances are produces or consumes in electrochemical reaction by given quantity of electric charge passed.

These laws have successfully summarize the stoichiometry of electrochemical. Not only stated the law Faraday also measured the charge mass ratio of the hydrogen ion, the value he obtained is about 1000 times smaller than e/m_e, then conclude that "the charge on the electron must be very large or the mass of the electron must be very small, compared to the hydrogen ion" .

The charge of one mole electron has been given a special name and symbol as Faraday constant, which has amount as, $F = 96{,}485.34\ C\ mol^{-1}$.

Chemical Kinetic

Chemical study major subject is reaction rates which involves the explanation of how do catalyst increase rates of chemical reaction. Why do small changes in temperature often have such large effect on the cooking rate of food. How does the way molecules combine to form of product and so on of this kind of question make this study becoming so complex.

The primary goal of chemical kinetics is to reduce the mechanism of reaction from experimental studies of its rates. The reaction mechanism itself is the slow or fast a reaction could produce its final products.

Rates of Chemical Reactions

The speed of reaction depends on many factors. Concentration of the reacting compound is the one who have significant effect. The other factor that effect is temperature which means that careful control of temperature is critical for quantitative measurement in chemical kinetic. Finally, the rate often depend crucially on the physical form if reactants.

A kinetics experiment involves measuring the rate of change of substance who participating in chemical reaction. A series of measurement which use different method for different condition could reveal the rate of change of the concentration. For low speed, we can use low temperature condition and in very fast reaction often used a flash of light whose rate is then tracked by measuring the absorption at particular wave length.

We conclude that once product from reaction is formed it can react back to give original reaction, the net rate is the difference which written in specific equation such as:

Net rate = forward rate – reverse rate

This means measurement of concentration give the net rate rather than simply the forward rate. In addition, many reactions go to completion mean they have a measurable rate only in forward direction or else the experiment can be arranged so that the product are removed as they are formed. The reaction between the rate of a reaction and concentration is called rate law.

Reaction Mechanism

Many reaction do not occur in single step, but rather than it sequence in steps to arrive at the product. Each of this step called as elementary reaction and occurs through the collision of atom, ions, or molecules. A unimolecular elementary reaction involves only a single reactant molecule. The example for this reaction is the dissociation of energized N_2O_3 molecules in gas phase. The most common type of elementary reaction involves the collision of atom, ions, or molecule which called biomolecular the example of this reaction is collision of NO + O_3 in gas phase which is the reverse reaction of N_2O_3.

The term termoecular is used for a reaction step that involves the simultaneous collision of three molecules, which is a much less most likely reaction for instance the recombination if iodine atoms in the gas phase to form iodine molecules. A reaction mechanism is a detailed sequence of elementary reactions, the details provide the rate of the molecule combined to yield the overall reaction. A reaction intermediate is a chemical species that is formed and consumed in the reaction but does not appear in the overall balanced chemical equations.

There is a direct relation between rates for elementary steps in a chemical reaction mechanism and the overall equilibrium constant K. This principle called detailed balance which states that at equilibrium the rate each elementary process is balanced by the rate of its reverse process. The prove of this principal is that rate constants of three steps elementary reaction are k_1,k,k_3 while the reverse of the three steps elementary reaction are k_{-1},k_{-2},k_{-3}.

Spectroscopy and Photochemistry

Spectroscopy

The key concept or the backbone of spectroscopy is quantization of energy. Energy quantization is a revolutionary concept that took most the early twentieth century to discover. Spectroscopy itself covers study about measures various form of light energy that are absorbed o emitted only at specific wave light.

Spectroscopy involves measurement of electromagnetic radiation as emitted from atoms and molecules between definite energy levels and can be characterized by wave length or frequency. Light energy is proportional to frequency with, $\varepsilon = hv$ and angular momentum is quantized as well ($mvr = nh$).

Photochemistry

Photochemistry is a study of chemical reaction produced by light, because the absorption of a photon from light a molecule may raise to an excited electronic state where it will be more likely to react than in the ground electronic state. In photochemical reactions, the activation energy is supplied by absorption of light. This concept is the contrary of thermal reaction that activation energy is supplied by intermolecular collisions.

The energy of photon can be determined by equation of $E_{photon} = hv = hc/\lambda$. So, the energy of one mole of photon is $N_A hv$. According to the Stark-Einstein law of photochemistry, the number of photon absorbed equals the number of molecules making transition to an excited electronic state. The term fluorescence does mean there is no change in spontaneous emission of radiation by an electronic transition in which the total electronic spin. In other hand phosphorescence mean emission that follow electronic excitation by absorption of light. And the term luminescence refers to any emission of light by electronically excited species, and includes fluorescence and phosphorescence.

The example of photochemistry that happen in plant generally in leaves specially is photosythesis which is the process where plant uses ultaviolet energy to convert dioxide and water into glucose and oxygen. The concept of photochemistry is about light, the photosythesis itself use light as it input and water to produce glucose and oxygen each in solid or fluid form and gas form.

Quantum Mechanic

Quantum mechanic provides a firm conceptual foundation for understanding several phenomena that chemist seek to understand. Such as, the atomic interaction that form molecule and extended solid structure, intermolecular interactions, and the chemical reactivity.

Quantum Chemistry

With a lot of development in quantum chemistry, today quantum chemistry provides very good approximate solutions to Schrodinger equation for molecules. It also provides a qualitative concept for representing and describing chemical bonds, molecular structure and chemical reactivity.

The Born- Oppenheimer approximation is the starting point for quantum mechanic. The points of Born- Oppenheimer approximation are nuclei are so much heavier than electrons that may be considered fixed space while the electron move rapidly around nuclei. The fact that electrons move so much faster than nuclei allows us to treat electronic motion independently of nuclear motion by solving an electronic Schrodinger equation for each value in inter nuclear separation.

Surface Chemistry

In these branches of physical chemistry, according to its title it means this study is about chemical reaction happen on the surfaces and interfaces. This study focusing on how molecule or atom interact with each other on the surface. The key point of studying molecule interaction is to understand the desirable chemical reactions for instance heterogeneous catalysis.

The purpose of studying the chemical reaction is in case to modify the chemical reaction so it can produce a various desired effect and or improvement locate on surface and interfaces. There are several fields who are fond of this study such as catalysis, electrochemistry, and geochemistry.

Meanwhile, there are sub branches of physical chemistry that used in daily life to compound many basic of industrial, laboratory, and many studies.

Henry's Law

Henry's Law states that the solubility of a gas in a liquid is directly proportional to the partial pressure of the gas above the liquid. Henry's law can be written as follows:

$$\frac{S_1}{P_1} = \frac{S_2}{P_2}$$

S_1 and P_1 are the solubility and the pressure at an initial set of conditions; S_2 and P_2 are the solubility and pressure at another changed set of conditions. Solubilities of gases are typically reported in g/L.

Pressure has very little effect on the solubility of solids or liquids, but has a significant effect on the solubility of gases. Gas solubility increases as the partial pressure of a gas above the liquid increases. Suppose a certain volume of water is in a closed container with the space above it occupied by carbon dioxide gas at standard pressure. Some of the CO_2 molecules come into contact with the surface of the water and dissolve into the liquid. Now suppose that more CO_2 is added to the space above the container, causing a pressure increase. More CO_2 molecules are now in contact with the water and so more of them dissolve. Thus the solubility increases as the pressure increases. As with a solid, the CO_2 that is undissolved reaches an equilibrium with the dissolved CO_2, represented by the following equation:

$$CO_2(g) \rightleftarrows CO_2(aq)$$

At equilibrium, the rate of gaseous CO_2 dissolving is equal to the rate of dissolved CO_2 coming out of the solution.

When carbonated beverages are packaged, they are done so under high CO_2 pressure so that a large amount of carbon dioxide dissolves in the liquid. When the bottle is open, the equilibrium is disrupted because the CO_2 pressure above the liquid decreases. Immediately, bubbles of CO_2 rapidly exit the solution and escape out of the top of the open bottle. The amount of dissolve CO_2 decreases. If the bottle is left open for an extended period of time, the beverage becomes "flat" as more and more CO_2 comes out of the liquid.

Henry's Law Constant

The mathematical expression of Henry's law is:

$$p = k_{\mathrm{H}} \cdot c$$

Above the solution, the solute's partial pressure is determined as p.

The solutes concentration in the solution is represented by c.

The Henry's Law Constant is k_{H}.

The various values of k_{H} are:

- 2 L · atm/mol: O_2 (oxygen)
- 4 L · atm/mol: Carbon Dioxide (CO_2)
- 1 L · atm/mol: Hydrogen (H_2)

This is all happening at a temperature of 298 K.

Temperature Dependency

The Henry's law of constant will change with the change in the temperature of the system. The temperature's effect on the constant is assessed by multiple equations. Some of the examples of the equations that are used to represent it are Van't Hoff equation:

$$k_{\mathrm{H,\,pc}} = k^{\ominus}_{\mathrm{H,\,pc}} \exp\left[-C \cdot \left(\frac{1}{T} - \frac{1}{T^{\ominus}}\right)\right]$$

$$k_{\mathrm{H,\,cp}} = k^{\ominus}_{\mathrm{H,\,cp}} \exp\left[C \cdot \left(\frac{1}{T} - \frac{1}{T^{\ominus}}\right)\right]$$

- The Henry's Law Constant is k_{H}.
- A constant that is measured in Kelvin's is C.
- Temperature is T (Kelvin).
- 298K is the standard state temperature or $T^{\ominus}$.

The equations mentioned above are used when a given gas has no other experimentally derived formula.

The phenomenon of decompression sickening is an instance, where Henry's law comes into play. When a person dives into the water and swims too deep, suffers the changes in compression that affect the oxygen and nitrogen content in the blood. A similar instance can be seen with soft drinks that have been carbonated by the use of carbon dioxide. The carbon dioxide is filled in the space above the drink, which has a higher pressure than the atmospheric pressure. The dissolved carbon dioxide undergoes degassing as the cap is opened and the partial pressure of carbon dioxide is lowered.

Hess's Law

Hess's Law of Constant Heat Summation (or just Hess's Law) states that regardless of the multiple stages or steps of a reaction, the total enthalpy change for the reaction is the sum of all changes. This law is a manifestation that enthalpy is a state function.

Hess's Law is named after Russian Chemist and Doctor Germain Hess. Hess helped formulate the early principles of thermochemistry. His most famous paper, which was published in 1840, included his law on thermochemistry. Hess's law is due to enthalpy being a state function, which allows us to calculate the overall change in enthalpy by simply summing up the changes for each step of the way, until product is formed. All steps have to proceed at the same temperature and the equations for the individual steps must balance out. The principle underlying Hess's law does not just apply to Enthalpy and can be used to calculate other state functions like changes in Gibbs' Energy and Entropy.

A pictorial view of Hess's Law as applied to the heat of equation $2H_2O_{(g)} \rightarrow 2H_{2(g)} + O_{2(g)}$ is illustrative. In figure, the reactants $C(s) + 2\,H_2O(g)$ are placed together in a box, representing the state of the materials involved in the reaction prior to the reaction. The products $CO_2(g) + 2\,H_2(g)$ are placed together in a second box representing the state of the materials involved after the reaction. The reaction arrow connecting these boxes is labeled with the heat of this reaction. Now we take these same materials and place them in a third box containing $C(s), O_2(g)$, and $2\,H_2(g)$. This box is connected to the reactant and product boxes with reaction arrows, labeled by the heats of reaction in equation $C_{(s)} + O_{2(g)} \rightarrow CO_{2(g)}$ and equation $2H_{2(g)} + O_{2(g)} \rightarrow 2H_2O_{(g)}$.

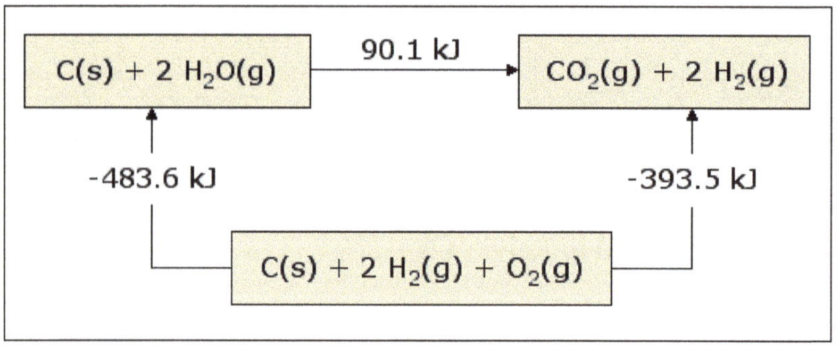

Figure: A Pictorial View of Hess's Law.

This picture of Hess's Law reveals that the heat of reaction along the "path" directly connecting the reactant state to the product state is exactly equal to the total heat of reaction along the alternative "path" connecting reactants to products via the intermediate state containing $C_{(s)}$, $O_{2(g)}$, and 2 $H_{2(g)}$. A consequence of our observation of Hess's Law is therefore that the net heat evolved or absorbed during a reaction is independent of the path connecting the reactant to product (this statement is again subject to our restriction that all reactions in the alternative path must occur under constant pressure conditions).

A slightly different view of figure 1 results from beginning at the reactant box and following a complete circuit through the other boxes leading back to the reactant box, summing the net heats of reaction as we go. We discover that the net heat transferred (again provided that all reactions occur under constant pressure) is exactly zero. This is a statement of the conservation of energy: the energy in the reactant state does not depend upon the processes which produced that state. Therefore, we cannot extract any energy from the reactants by a process which simply recreates the reactants. Were this not the case, we could endlessly produce unlimited quantities of energy by following the circuitous path which continually reproduces the initial reactants.

By this reasoning, we can define an energy function whose value for the reactants is independent of how the reactant state was prepared. Likewise, the value of this energy function in the product state is independent of how the products are prepared. We choose this function, H, so that the change in the function, $\Delta H = H_{products} - H_{reactants}$, is equal to the heat of reaction q under constant pressure conditions. H, which we call the enthalpy, is a state function, since its value depends only on the state of the materials under consideration, that is, the temperature, pressure and composition of these materials.

The concept of a state function is somewhat analogous to the idea of elevation. Consider the difference in elevation between the first floor and the third floor of a building. This difference is independent of the path we choose to get from the first floor to the third floor. We can simply climb up two flights of stairs, or we can climb one flight of stairs, walk the length of the building, then walk a second flight of stairs. Or we can ride the elevator. We could even walk outside and have a crane lift us to the roof of the building, from which we climb down to the third floor. Each path produces exactly the same elevation gain, even though the distance traveled is significantly different from one path to the next. This is simply because the elevation is a "state function". Our elevation, standing on the third floor, is independent of how we got to the third floor, and the same is true of the first floor. Since the elevation thus a state function, the elevation gain is independent of the path. Now, the existence of an energy state function H is of considerable importance in calculating heats of reaction. Consider the prototypical reaction in figure, with reactants R being converted to products P. We wish to calculate the heat absorbed or released in this reaction, which is ΔH. Since H is a state function, we can follow any path from R to P and calculate ΔH along that path. In figure below, we consider one such possible path, consisting of two reactions passing through an intermediate state containing all the atoms involved in the reaction, each in elemental form. This is a useful intermediate state since it can be used for any possible chemical reaction. For example, the atoms involved in the reaction are C, H, and O, each of which are represented in the intermediate state in elemental form. We can see in figure that the ΔH for the overall reaction is now the difference between the ΔH in the formation of the products P from the elements and the ΔH in the formation of the reactants R from the elements.

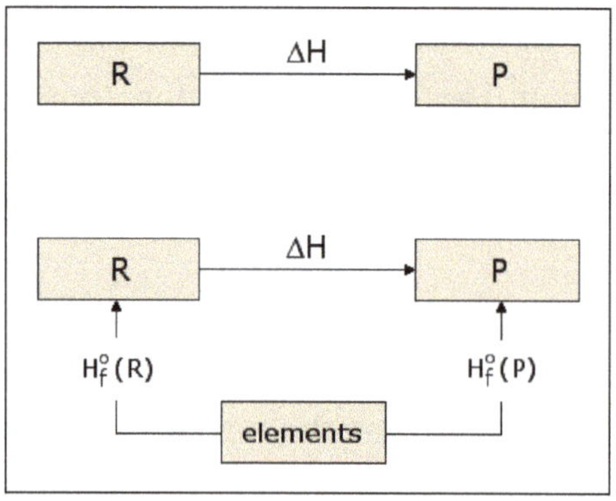

Figure: Calculation of ΔH.

The ΔH values for formation of each material from the elements are thus of general utility in calculating ΔH for any reaction of interest. We therefore define the standard formation reaction for reactant R, as elements in standard state → R and the heat involved in this reaction is the standard enthalpy of formation, designated by ΔH_f°. The subscript f, standing for "formation," indicates that the ΔH is for the reaction creating the material from the elements in standard state. The superscript ∘ indicates that the reactions occur under constant standard pressure conditions of 1 atm. From figure, we see that the heat of any reaction can be calculated from:

$$\Delta H_f^{\circ} = \Delta H^{\circ}_{f,products} - \Delta H^{\circ}_{f,\ reactants}$$

Extensive tables of ΔH_f° values have been compiled that allows us to calculate with complete confidence the heat of reaction for any reaction of interest, even including hypothetical reactions which may be difficult to perform or impossibly slow to react.

Example: The enthalpy of a reaction does not depend on the elementary steps, but on the final state of the products and initial state of the reactants. Enthalpy is an extensive property and hence changes when the size of the sample changes. This means that the enthalpy of the reaction scales proportionally to the moles used in the reaction. For instance, in the following reaction, one can see that doubling the molar amounts simply doubles the enthalpy of the reaction.

$$H_2(g) + 1/2O_2(g) \rightarrow H_2O(g)\ \Delta H^\circ = -572\,kJ$$

$$2H_2(g) + O_2(g) \rightarrow 2H_2O(g)\ \Delta H^\circ = -1144\,kJ$$

The sign of the reaction enthalpy changes when a process is reversed.

$$H_2(g) + 1/2O_2(g) \rightarrow H_2O(g)\ \Delta H^\circ = -572\,kJ$$

When switched:

$$H_2O(g) \rightarrow H_2(g) + 1/2O_2(g)\ \Delta H^\circ = +572\,kJ$$

Since enthalpy is a state function, it is path independent. Therefore, it does not matter what reactions one uses to obtain the final reaction.

Steps involved in solving enthalpy of combustion problems:

- Balance the individual equations.
- If necessary look up standard enthalpies.
- Flip equations around if necessary to cancel out terms on opposite sides.
- Changing the equation around requires a sign change of the H of that individual step.
- Sum up the individual steps.

Example of Graphite to Diamond

Calculate the standard enthalpy of combustion of the transition of $C_{(s,\ graphite)} \rightarrow C_{(s,\ diamond)}$, given:

$$C_{(s,\ graphite)} + O_2 \rightarrow CO_2 \quad \Delta H^\circ = -393.5 kJ / mol$$

$$CO_2 \rightarrow C_{(s,\ diamond)} + O_2 \quad \Delta H^\circ = +395.41 kJ / mol$$

First we see that both equations are balanced. The enthalpies were given and there is no need to flip an equation around because it is possible to cancel out a couple terms as is. What is left is canceling out the O_2 and the CO_2 species, writing the overall reaction and then summing the two enthalpies together:

$$C_{(s,\ graphite)} + \cancel{O_2} \rightarrow \cancel{CO_2}$$

$$\cancel{CO_2} \rightarrow C_{(s,\ diamond)} + \cancel{O_2}$$

Overall equation becomes $C_{(s,\ graphite)} \rightarrow C_{(s,\ diamond)}$

Adding the enthalpies gives (-393.5 kJ/mol + 395.41 kJ/mol) = +1.91 kJ/mol. Since the ΔH^o is positive, the reaction is endothermic.

Standard Enthalpy of Formation

The following equation is used to calculate the Standard Enthalpy of Formation:

$$\Delta_r H^\ominus = \sum \upsilon H^\ominus(\text{products}) - \sum \upsilon H^\ominus(\text{reactants})$$

- If ΔH^o is positive, the reaction is endothermic.
- If ΔH^o is negative, the reaction is exothermic.

Standard Gibb's Energy of Formation

As mentioned above, the Standard Gibb's Free Energy of Formation is calculated using the principles of Hess's law. The following equation is used for those calculations:

$$\Delta_f G^\ominus = \sum v \Delta_f G^\ominus(products) - \sum v \Delta_f G^\ominus(reactants)$$

Standard Entropy

The Standard Entropy of a reaction is calculated using the following equation:

$$\Delta_r S^{\ominus} = \sum \upsilon S^{\ominus}(products) - \sum \upsilon S^{\ominus}(reactants)$$

Note that in contrast to Gibbs Energy and Enthalpy, the third law of thermodynamics lets us work in absolute (SS) instead of differences with respect to a reference (standard) state.

Example of Steam Reforming

Steam reforming is a method for producing hydrogen, carbon monoxide, or other useful products from hydrocarbon fuels such as natural gas. Use Hess's Law to calculate the following enthalpy of reaction of the major process of steam reforming:

$$CH_{4(g)} + H_2O_{(g)} \rightleftharpoons CO_{(g)} + 3H_{2(g)}$$

Given the separate reactions of carbon dioxide and hydrogen gas and methane decomposition:

$$CO_{(g)} + H_{2(g)} \rightarrow C_{(graphite)} + H_2O_{(g)} \quad \Delta H° = -131.3\ kJ$$

$$C_{(graphite)} + 2H_{2(g)} \rightarrow CH_{4(g)} \quad \Delta H^{\circ} = -74.8\ kJ$$

Using the two postulates, given enough information, we are able to solve the enthalpy of reaction of an untabulated equation:

1. Flip equation $C_{(graphite)} + 2H_{2(g)} \rightarrow CH_{4(g)} \quad \Delta H^{\circ} = -74.8\ kJ$ to and change the sign of the enthalpy reaction.

   $$CH_{4(g)} \rightarrow C_{(graphite)} + 2H_{2(g)} \quad \Delta H° = +74.8\ kJ$$

2. Flip equation $CO_{(g)} + H_{2(g)} \rightarrow C_{(graphite)} + H_2O_{(g)} \quad \Delta H° = -131.3\ kJ$ as well in order to get the unnecessary equation parts to cancel out.

   $$C_{(graphite)} + H_2O \rightarrow CO_{(g)} + H_{2(g)} \quad \Delta H° = +131.3\ kJ$$

3. With the two equations packed together, notice the equations will cancel out and come out with the final equation. C(graphite's) cancel out between the two equations.

   $$\cancel{C_{(graphite)}} + H_2O \rightarrow CO_{(g)} + H_{2(g)} \quad \Delta H° = +131.3\ kJ$$

   $$CH_{4(g)} \rightarrow \cancel{C_{(graphite)}} + 2H_{2(g)} \quad \Delta H° = +74.8\ kJ$$

   Final equation:

   $$CH_{4(g)} + H_2O \rightarrow CO + 3H_2 \quad \Delta H\circ = +74.8\ kJ + 131.3\ kJ$$

4. Add the final enthalpies to the two fixed up equations and receive the enthalpy of the unknown equation ΔH° = +201.1 kJ:

$$CH_{4(g)} + H_2O \rightarrow CO + 3H_2 \quad \Delta H^\circ = 201.1\ kJ$$

Hence, it is not surprizing that this reforming reactions takes place at high temperatures.

Application

Hydrogen gas, which is of potential interest nationally as a clean fuel, can be generated by the reaction of carbon (coal) and water:

$$C_{(s)} + 2H_2O_{(g)} \rightarrow CO_{2(g)} + 2H_{2(g)}$$

Calorimetry reveals that this reaction requires the input of 90.1 kJ of heat for every mole of $C_{(s)}$ consumed. By convention, when heat is absorbed during a reaction, we consider the quantity of heat to be a positive number: in chemical terms, $q > 0$ for an endothermic reaction. When heat is evolved, the reaction is exothermic and $q < 0$ by convention.

It is interesting to ask where this input energy goes when the reaction occurs. One way to answer this question is to consider the fact that the reaction converts one fuel, $C_{(s)}$, into another, $H_{2(g)}$. To compare the energy available in each fuel, we can measure the heat evolved in the combustion of each fuel with one mole of oxygen gas. We observe that,

$$C_{(s)} + O_{2(g)} \rightarrow CO_{2(g)}$$

produces 393.5 kJ for one mole of carbon burned; hence q = −393.5 kJ.

$$2H_{2(g)} + O_{2(g)} \rightarrow 2H_2O_{(g)}$$

The reaction produces 483.6 kJ for two moles of hydrogen gas burned, so q = -483.6 kJ. It is evident that more energy is available from combustion of the hydrogen fuel than from combustion of the carbon fuel, so it is not surprising that conversion of the carbon fuel to hydrogen fuel requires the input of energy. Of considerable importance is the observation that the heat input in equation $C_{(s)} + 2H_2O_{(g)} \rightarrow CO_{2(g)} + 2H_{2(g)}$, 90.1 kJ, is exactly equal to the difference between the heat evolved, -393.5 kJ, in the combustion of carbon and the heat evolved, -483.6 kJ, in the combustion of hydrogen. This is not a coincidence: if we take the combustion of carbon and add to it the reverse of the combustion of hydrogen, we get

$$C_{(s)} + O_{2(g)} \rightarrow CO_{2(g)}$$

$$2H_2O_{(g)} \rightarrow 2H_{2(g)} + O_{2(g)}$$

$$C_{(s)} + O_{2(g)} + 2H_2O_{(g)} \rightarrow CO_{2(g)} + 2H_{2(g)} + O_{2(g)}$$

Canceling the $O_{2(g)}$ from both sides, since it is net neither a reactant nor product, equation $C_{(s)} + O_{2(g)} + 2H_2O_{(g)} \rightarrow CO_{2(g)} + 2H_{2(g)} + O_{2(g)}$ is equivalent to equation $2H_2O_{(g)} \rightarrow 2H_{2(g)} + O_{2(g)}$.

Thus, taking the combustion of carbon and "subtracting" the combustion of hydrogen (or more accurately, adding the reverse of the combustion of hydrogen) yields equation $2H_2O_{(g)} \rightarrow 2H_{2(g)} + O_{2(g)}$. And, the heat of the combustion of carbon minus the heat of the combustion of hydrogen equals the heat of equation $2H_2O_{(g)} \rightarrow 2H_{2(g)} + O_{2(g)}$. By studying many chemical reactions in this way, we discover that this result, known as Hess's Law, is general.

Raoult's Law

Raoult's law is a chemical law that relates the vapor pressure of a solution is dependent on the mole fraction of a solute added to solution.

Raoult's Law is expressed by the formula:

$$P_{solution} = X_{solvent} P^0_{solvent}$$

where:

- $P_{solution}$ is the vapor pressure of the solution.
- $X_{solvent}$ is mole fraction of the solvent.
- $P^0_{solvent}$ is the vapor pressure of the pure solvent.

If more than one solute is added to the solution, each individual solvent's component is added to the total pressure.

Raoult's law is akin to the ideal gas law, except for solution. The ideal gas law assumes ideal behavior in which the intermolecular forces between dissimilar molecules equals forces between similar molecules. Raoult's law assumes the physical properties of the components of a chemical solution are identical.

Deviations from Raoult's Law

If there are adhesive or cohesive forces between two liquids, there will be deviations from Raoult's law. There is negative deviation when the vapor pressure is lower than expected from the law. This occurs when forces between particles are stronger than those between particles in pure liquids. This behavior is observed in a mixture of chloroform and acetone. Here, hydrogen bonds cause the deviation. Another example of negative deviation is in a solution of hydrochloric acid and water.

Positive deviation occurs when the cohesion between similar molecules exceeds adhesion between unlike molecules. The result is higher than expected vapor pressure. Both components of the mixture escape solution more readily than if the components were pure. This behavior is observed in mixtures of benzene and methanol and mixtures of chloroform and ethanol.

Ideal vs. Nonideal Solutions

Raoult's Law only works for ideal solutions. "An ideal solution shows thermodynamic mixing characteristics identical to those of ideal gas mixtures except ideal solutions have intermolecular interactions equal to those of the pure components." Like many other concepts explored in Chemistry, Raoult's Law only applies under ideal conditions in an ideal solution. However, it still works fairly well for the solvent in dilute solutions. In reality though, the decrease in vapor pressure will be greater than that calculated by Raoult's Law for extremely dilute solutions.

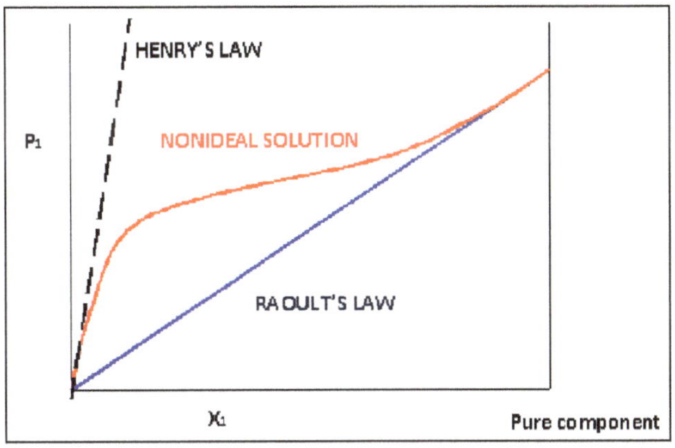

Figure: Positive non-ideal behavior of the vapor pressure of a solution follows Henry's Law at low concentrations and Raoult's Law at high concentrations (pure).

Why Raoult's Law Works?

If you look review the concepts of colligative properties, you will find that adding a solute lowers vapor pressure because the additional solute particles will fill the gaps between the solvent particles and take up space. This means less of the solvent will be on the surface and less will be able to break free to enter the gas phase, resulting in a lower vapor pressure. There are two ways of explaining why Raoult's Law works - a simple visual way, and a more sophisticated way based on entropy. Below is the simple approach.

Remember that saturated vapor pressure is what you get when a liquid is in a sealed container. An equilibrium is set up where the number of particles breaking away from the surface is exactly the same as the number sticking on to the surface again.

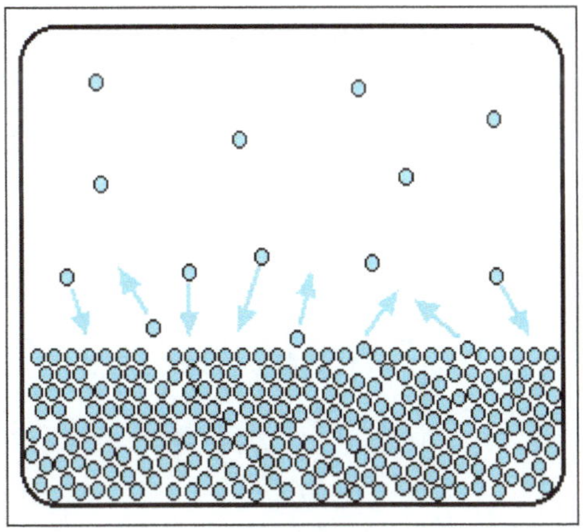

Figure: Dynamic equilibrium between volatile molecules in the liquid and gas phase.

Now suppose solute molecules were added so that the solvent molecules occupied only 50% of the surface of the solution.

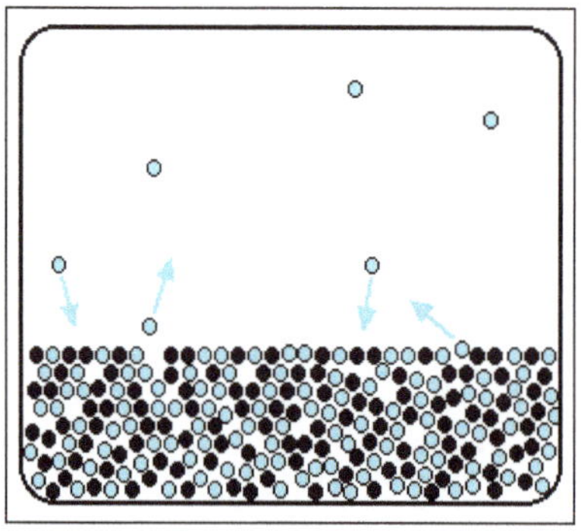

Figure: The Dynamic equilibrium of a solution consisting of volatile solvent molecules (teal circles) and non-volatile solute molecules (black circles).

A certain fraction of the solvent molecules will have sufficient energy to escape from the surface (e.g., 1 in 1000 or 1 in a million). If you reduce the number of solvent molecules on the surface, you are going to reduce the number which can escape in any given time. But it will not make any difference to the ability of molecules in the vapor to stick to the surface again. If a solvent molecule in the vapor hits a bit of surface occupied by the solute particles, it may well stick. There are obviously attractions between solvent and solute otherwise you would not have a solution in the first place.

The net effect of this is that when equilibrium is established, there will be *fewer* solvent molecules in the vapor phase - it is less likely that they are going to break away, but there is not any problem about them returning. However, if there are fewer particles in the vapor at equilibrium, the saturated vapor pressure is lower.

Limitations on Raoult's Law

In practice, there's no such thing as an ideal solution. However, features of one include:

- Ideal solutions satisfy Raoult's Law. The solution in the figure above would not actually obey Raoult's Law - it is far too concentrated, but was drawn so concentrated to emphasized the point.
- In an ideal solution, it takes exactly the same amount of energy for a solvent molecule to break away from the surface of the solution as it did in the pure solvent. The forces of attraction between solvent and solute are exactly the same as between the original solvent molecules - not a very likely event.

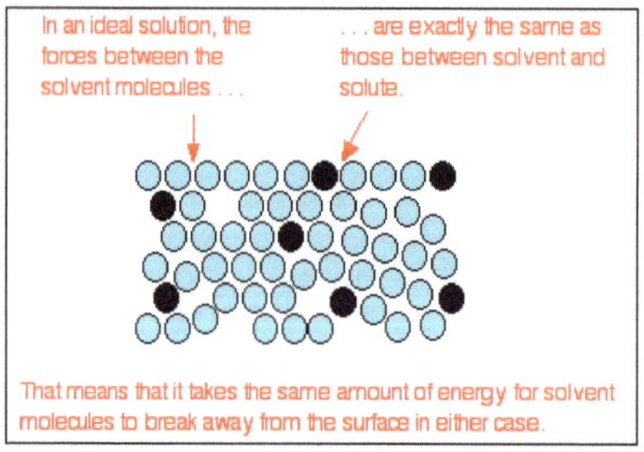

Suppose that in the pure solvent, 1 in 1000 molecules had enough energy to overcome the intermolecular forces and break away from the surface in any given time. In an ideal solution, that would still be exactly the same proportion. Fewer would, of course, break away because there are now fewer solvent molecules on the surface - but of those that are on the surface, the same proportion still break away. If there were strong solvent-solute attractions, this proportion may be reduced to 1 in 2000, or 1 in 5000 or whatever.

In any real solution of, say, a salt in water, there are strong attractions between the water molecules and the ions. That would tend to slow down the loss of water molecules from the surface. However, if the solution is sufficiently dilute, there will be good-sized regions on the surface where you still have water molecules on their own. The solution will then approach ideal behavior.

The Nature of the Solute

There is another thing that you have to be careful of if you are going to do any calculations on Raoult's Law. You may have noticed in the little calculation about mole fraction, that sugar was as a solute rather than salt. For each mole of sodium chloride dissolved, you get 1 mole of sodium ions and 1 mole of chloride ions - in other words, you get twice the number of moles of particles as of original salt.

$$\underset{\text{1 mole of solid salt}}{Na^{+}Cl^{-}_{(s)}} \longrightarrow \underset{\text{2 moles of ions in solution}}{Na^{+}_{(aq)} + Cl^{-}_{(aq)}}$$

So, if you added 0.1 moles of sodium chloride, there would actually be 0.2 moles of particles in the solution - and that's the figure you would have to use in the mole fraction calculation. Unless you think carefully about it, Raoult's Law only works for solutes which do not change their nature when they dissolve. For example, they must not ionize or associate (e.g., if you put in substance A, it must not form A_2 in solution). If it does either of these things, you have to treat Raoult's law with great care.

Raoult's Law and Colligative Properties

The effect of Raoult's Law is that the saturated vapor pressure of a solution is going to be lower than that of the pure solvent at any particular temperature. That has important effects on the phase diagram of the solvent. The next diagram shows the phase diagram for pure water in the region around its normal melting and boiling points. The 1 atmosphere line shows the conditions for measuring the normal melting and boiling points.

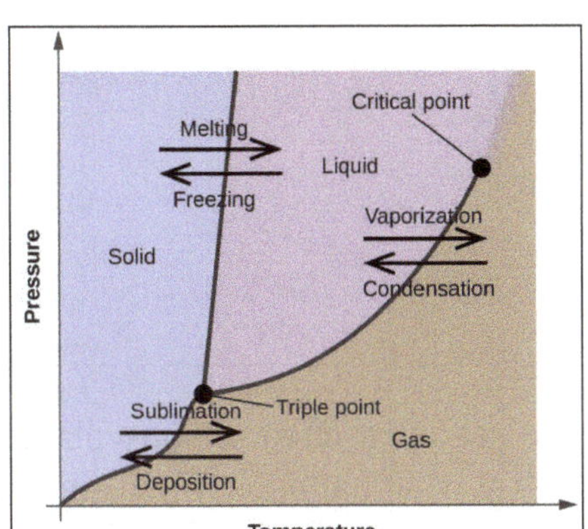

Figure: Phase diagram of a pure solvent.

The line separating the liquid and vapor regions is the set of conditions where liquid and vapor are in equilibrium. It can be thought of as the effect of pressure on the boiling point of the water, but it is also the curve showing the effect of temperature on the saturated vapor pressure of the water.

If you draw the saturated vapor pressure curve for a solution of a non-volatile solute in water, it will always be lower than the curve for the pure water.

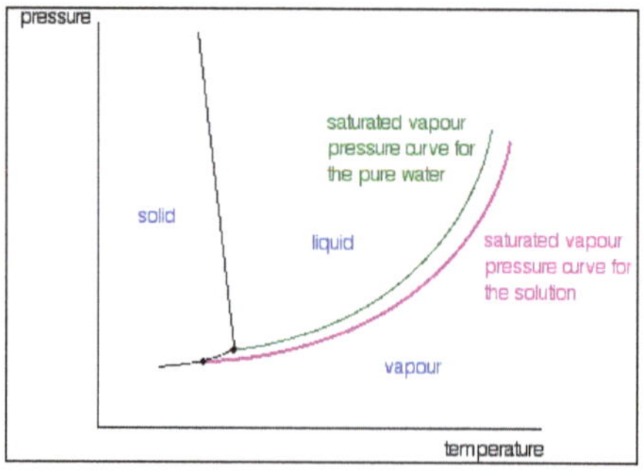

If you look closely at the last diagram, you will see that the point at which the liquid-vapor equilibrium curve meets the solid-vapor curve has moved. That point is the triple point of the system - a unique set of temperature and pressure conditions at which it is possible to get solid, liquid and vapor all in equilibrium with each other at the same time. Since the triple point has solid-liquid equilibrium present (amongst other equilibria), it is also a melting point of the system - although not the normal melting point because the pressure is not one atmosphere.

The curves for the pure water and for the solution are often drawn parallel to each other. Suppose you have a solution where the mole fraction of the water is 0.99 and the vapor pressure of the pure water at that temperature is 100 kPa. The vapor pressure of the solution will be 99 kPa - a fall of 1 kPa. At a lower temperature, where the vapor pressure of the pure water is 10 kPa, the fall will only be 0.1 kPa. For the curves to be parallel the falls would have to be the same over the whole temperature range.

That must mean that the phase diagram needs a new melting point line (a solid-liquid equilibrium line) passing through the new triple point. That is shown in the next diagram.

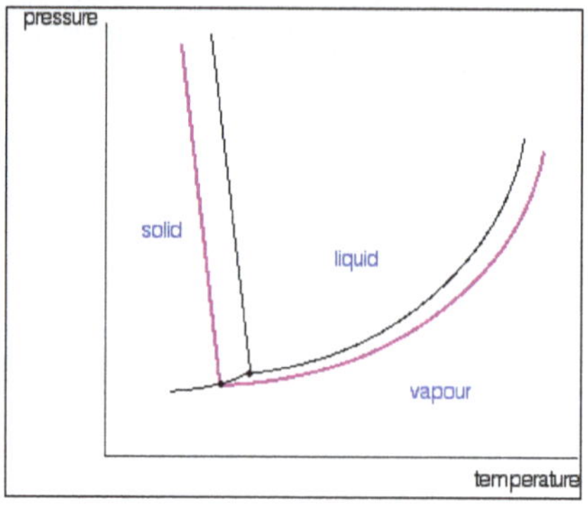

Now we are finally in a position to see what effect a non-volatile solute has on the melting and freezing points of the solution. Look at what happens when you draw in the 1 atmosphere pressure line which lets you measure the melting and boiling points. The diagram also includes the melting and boiling points of the pure water from the original phase diagram for pure water (black lines).

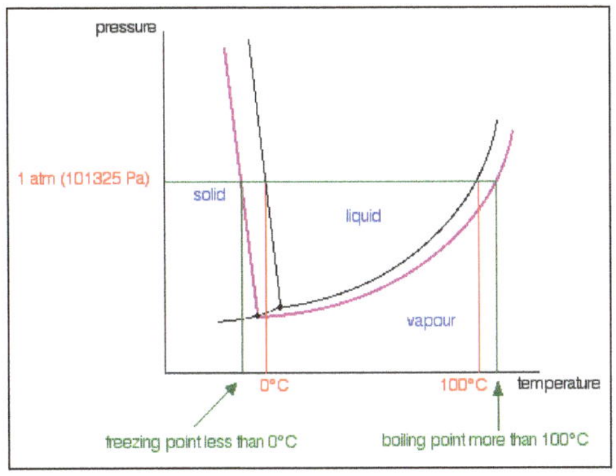

Because of the changes to the phase diagram, you can see that:

- The boiling point of the solvent in a solution is higher than that of the pure solvent.
- The freezing point (melting point) of the solvent in a solution is lower than that of the pure solvent.

We have looked at this with water as the solvent, but using a different solvent would make no difference to the argument or the conclusions. The only difference is in the slope of the solid-liquid equilibrium lines. For most solvents, these slope forwards whereas the water line slopes backwards. You could prove to yourself that that does not affect what we have been looking at by re-drawing all these diagrams with the slope of that particular line changed. You will find it makes no difference whatsoever.

How to Calculate the Vapor Pressure of a Solution

We can calculate the vapor pressure of the solution in two ways, depending on the volatility of the solute. If the solute is volatile, it will exert its own vapor pressure and this amount is a significant contribution to the overall vapor pressure of the solution, and thus needs to be included in the calculations. On the other hand, if it is nonvolatile, the solute will not produce vapor pressure in solution at that temperature.

Nonvolatile Solutes

These calculations are fairly straightforward if you are comfortable with stoichiometric conversions. Because the solute is nonvolatile, you need only determine the change in vapor pressure for the solvent. Using the equation for Raoult's Law, you will need to find the mole fraction of the solvent and the vapor pressure of the pure solvent is typically given.

Example of Kool-Aid

1.5 moles of cherry Kool-Aid are added to a pitcher containing 2 liters of water on a nice day at 25 °C. The vapor pressure of water alone is 23.8 mm Hg at 25 °C. What is the new vapor pressure of Kool-Aid?

Solution:

$P_{H_2O} = 23.8$ mm Hg

To solve for the mole fraction, you must first convert the 2 L of water into moles:

1 L = 1000 mL = 1000 g

Knowing this, you can convert the mass of water (2000 g) into moles:

2000 g / 18.02 g (molar mass of water) = 110.9 moles H_2O

Solve for the mole fraction, χH_2O:

χ_{H_2O} = moles H_2O / total moles

= 110.9 moles / 110.9 + 1.5 moles = 0.979

Finally, apply Raoult's Law:

$P_{Kool-Aid} = \chi_{H_2O} P_{H_2O}$ = (0.979)(23.8 mm Hg) = 23.3 mm Hg

Example of Non-electrolyte

Calculate the vapor pressure of a solution made by dissolving 50.0 g glucose, $C_6H_{12}O_6$, in 500 g of water. The vapor pressure of pure water is 47.1 torr at 37 °C.

Solution:

To use Raoult's Law ($P_{solution} = \chi_{solvent} P^o_{solvent}$), we need to calculate the mole fraction of water (the solvent) in this sugar-water solution.

$$\chi_{solvent} = \frac{\text{moles of water}}{\text{moles of solute} + \text{moles of solvent}}$$

$$\chi_{solvent} = \frac{n_{water}}{n_{glucose} + n_{water}}$$

The molar mass of glucose if 180.2 g/mol and of water is 18 g/mol. So,

$$n_{water} = \frac{500g}{18g/mol} = 27.7mol$$

and,

$$n_{glucose} = \frac{50g}{180.2g/mol} = 0.277mol$$

and,

$$\chi_{solvent} = \frac{27.7mol}{0.277mol + 27.7mol} = 0.99$$

Note that this still relatively dilute.

$$P_{solution} = 0.99 \times 47.1 = 46.63 \textit{ torr}$$

Example of Electrolyte

Calculate the vapor pressure of a solution made by dissolving 50.0 g $CaCl_2$, $C_6H_{12}O_6$, in 500 g of water. The vapor pressure of pure water is 47.1 torr at 37 °C.

Solution:

To use Raoult's Law ($P_{solution} =_{\chi solvent} P^o_{solvent}$), we need to calculate the mole fraction of water (the solvent) in this salt-water solution.

$$\chi_{solvent} = \frac{\text{moles of water}}{\text{moles of solute} + \text{moles of solvent}}$$

$$\chi_{solvent} = \frac{n_{water}}{n_{solutes} + n_{water}}$$

The molar mass of $CaCl_2$ if 111 g/mol and of water is 18 g/mol. So,

$$n_{water} = \frac{500g}{18g/mol} = 27.7mol$$

and,

$$n_{solutes} = \frac{50g}{111g/mol} = 0.45mol$$

but this is really:

- $n_{Ca+} = 0.45mol$
- $n_{Cl-} = 0.9mol$

and,

$$\chi_{solvent} = \frac{27.7mol}{0.45mol + 0.9mol + 27.7mol} = 0.953$$

Note that this still relatively dilute.

$$P_{solution} = 0.953 \times 47.1 = 44.88torr$$

Volatile Solutes

The only difference between volatile and nonvolatile solutes, is that the partial pressure exerted by the vapor pressure of the volatile solute and the vapor pressure of the solvent must be accounted for. The sum of the two will give you the total vapor pressure of the solution.

Example: What are the partial pressures of benzene and toluene in a solution in which the mole fraction of benzene is 0.6? What is the total vapor pressure? The vapor pressure of pure benzene is 95.1 mm Hg and the vapor pressure of pure toluene 28.4 mm Hg at 25 °C.

Solution:

If $\chi_{benzene} = 0.6$, than $\chi_{toluene} = 0.4$ because 1−0.6=0.4.

Now that we know the mole fractions and vapor pressures, this problem is a cinch.

$$P_{benzene} = x_{benzene} P_{benzene} = (0.6)(95.1\,\text{mm Hg}) = 57.1\ \text{mm Hg}$$
$$P_{toluene} = x_{toluene} P_{toluene} = (0.4)(28.4\,\text{mm Hg}) = 11.4\,\text{mm Hg}$$

The total vapor pressure is simply the sum of the partial pressures:

$$P_{total} = P_{benzene} + P_{tuolene} = 57.1\,\text{mm Hg} + 11.4\,\text{mm Hg} = 68.5\,\text{mm Hg}.$$

Chemical Kinetics

Chemical kinetics is the branch of physical chemistry that is concerned with understanding the rates of chemical reactions. It is to be contrasted with thermodynamics, which deals with the direction in which a process occurs but in itself tells nothing about its rate. Thermodynamics is time's arrow, while chemical kinetics is time's clock. Chemical kinetics relates to many aspects of cosmology, geology, biology, engineering, and even psychology and thus has far-reaching implications. The principles of chemical kinetics apply to purely physical processes as well as to chemical reactions.

One reason for the importance of kinetics is that it provides evidence for the mechanisms of chemical processes. Besides being of intrinsic scientific interest, knowledge of reaction mechanisms is of practical use in deciding what is the most effective way of causing a reaction to occur. Many commercial processes can take place by alternative reaction paths, and knowledge of the mechanisms makes it possible to choose reaction conditions that favour one path over others.

A chemical reaction is, by definition, one in which chemical substances are transformed into other substances, which means that chemical bonds are broken and formed so that there are changes in the relative positions of atoms in molecules. At the same time, there are shifts in the arrangements of the electrons that form the chemical bonds. A description of a reaction mechanism must therefore deal with the movements and speeds of atoms and electrons. The detailed mechanism by which a chemical process occurs is referred to as the reaction path, or pathway.

The vast amount of work done in chemical kinetics has led to the conclusion that some chemical reactions go in a single step; these are known as elementary reactions. Other reactions go in more than one step and are said to be stepwise, composite, or complex. Measurements of the rates of chemical reactions over a range of conditions can show whether a reaction proceeds by one or more steps. If a reaction is stepwise, kinetic measurements provide evidence for the mechanism of the individual elementary steps. Information about reaction mechanisms is also provided by certain nonkinetic studies, but little can be known about a mechanism until its kinetics has been investigated. Even then, some doubt must always remain about a reaction mechanism. An investigation, kinetic or otherwise, can disprove a mechanism but can never establish it with absolute certainty.

Reaction Rate

The rate of a reaction is defined in terms of the rates with which the products are formed and the reactants (the reacting substances) are consumed. For chemical systems it is usual to deal with the concentrations of substances, which is defined as the amount of substance per unit volume. The rate can then be defined as the concentration of a substance that is consumed or produced in unit time. Sometimes it is more convenient to express rates as numbers of molecules formed or consumed in unit time.

The half-life

A useful rate measure is the half-life of a reactant, which is defined as the time that it takes for half of the initial amount to undergo reaction. For a special type of kinetic behaviour, the half-life is independent of the initial amount. A common and straightforward example of a half-life independent of the initial amount is radioactive substances. Uranium-238, for example, decays with a half-life of 4.5 billion years; of an initial amount of uranium, half of that amount will have decayed in that period of time. The same behaviour is found in many chemical reactions.

Even when the half-life of a reaction varies with the initial conditions, it is often convenient to quote a half-life, bearing in mind that it applies only to the particular initial conditions. Consider, for example, the reaction in which hydrogen and oxygen gases combine to form water; the chemical equation is:

$$2H_2 + O_2 \rightarrow 2H_2O.$$

If the gases are mixed together at atmospheric pressure and room temperature, nothing observable will happen over long periods of time. However, reaction does occur, with a half-life that is estimated to be more than 12 billion years, which is roughly the age of the universe. If a spark is passed through the system, the reaction occurs with explosive violence, with a half-life of less than one-millionth of a second. This is a striking example of the great range of rates with which chemical kinetics is concerned. There are many possible processes that proceed too slowly to be studied experimentally, but sometimes they can be accelerated, often by the addition of a substance known as a catalyst. Some reactions are even faster than the hydrogen-oxygen explosion—for example, the combination of atoms or molecular fragments (called free radicals) where all that occurs is the formation of a chemical bond. Some modern kinetic investigations are concerned with even faster processes, such as the breakdown of highly energetic and therefore transient molecules, where times of the order of femtoseconds (fs; 1 fs = 10^{-15} second) are involved.

Measuring Slow Reactions

The best way to study exceedingly slow reactions is to change the conditions so that the reactions occur in a reasonable time. Increasing the temperature, which can have a strong effect on the reaction rate, is one possibility. If the temperature of a hydrogen-oxygen mixture is raised to about 500 °C (900 °F), reaction then occurs rapidly, and its kinetics has been studied under those conditions. When a reaction occurs to a measurable extent over a period of minutes, hours, or days, rate measurements are straightforward. Amounts of reactants or products are measured at various times, and the rates are readily calculated from the results. Many automated systems have now been devised for measuring rates in this way.

Measuring Fast Reactions

Some processes are so fast that special techniques have to be used to study them. There are two difficulties with fast reactions. One is that the time that it takes to mix reactants or to change the temperature of the system may be significant in comparison with the half-life, so that the initial time cannot be measured accurately. The other difficulty is that the time it takes to measure the amounts of substances may be comparable with the half-life of the reaction. The methods used to overcome these difficulties fall into two classes: flow methods and pulse and probe methods.

In flow methods, two gases or solutions are introduced rapidly into a mixing vessel, and the resulting mixture then flows rapidly along a tube. Concentrations of reactants or products may then be measured—for example, by spectroscopic methods—at various positions along the tube, which correspond to various reaction times. A modification of this method is the stopped-flow technique, in which the reactants are forced rapidly into a reaction chamber; the flow is then suddenly stopped, and the amounts are measured by physical methods after various short times. These flow methods are limited by the time it takes to mix gases or solutions and are not suitable if the half-life is less than about a hundredth of a second.

These mixing difficulties were overcome by pulse and probe methods. The principle of these is that a short pulse, usually of radiation, is given to a chemical system and is then followed by a probe, usually involving radiation that provides spectroscopic evidence of what occurred after the initial pulse. The first of these methods, developed in 1949 by British chemists R.G.W. Norrish and George Porter, was the flash-photolysis method, for which Norrish and Porter won the Nobel Prize for Chemistry in 1967. In this technique a flash of light of high intensity but short duration brings about the formation of atomic and molecular species, the reactions of which can be studied kinetically by spectroscopy. In the earliest experiments the duration of the flash was about a millisecond (ms; 1 ms = 10^{-3} second), but in the next four decades the duration was reduced by more than 11 powers of 10, to just a few femtoseconds. A nanosecond (ns; 1 ns = 10^{-9} second) flash is adequate for studying almost any purely chemical reaction where there is a change in chemical identity. Any chemical reaction, however, involves processes of a purely physical nature, such as energy redistribution and the breakdown of transient species, which occur in the femtosecond range.

Many such processes have now been studied with flashes of only a few femtoseconds' duration. The time that it takes for the length of a chemical bond to change by 10^{-10} metre can be as little as about 100 fs, so that a flash of a few femtoseconds' duration, closely followed by another one of the same duration, will provide information about such tiny changes in bond lengths. The technique for causing one flash to occur a few nanoseconds after another is to route the light by a slightly longer path. A path of 1 additional micrometre (μm; 1 μm = 10^{-6} metre) causes a delay of 1 fs, and such a short path difference is now technically feasible. Egyptian-born chemist Ahmed Zewail won the Nobel Prize for Chemistry in 1999 for his work in this field.

Another pulse method is the relaxation method, developed in the 1950s by German physicist Manfred Eigen (who shared the Nobel Prize for Chemistry in 1967 with Norrish and Porter). In this method the investigation begins with a reaction system in equilibrium; the reaction to be studied has finished, and no further changes take place. The external conditions are then altered very rapidly; the system is then no longer at equilibrium, and it relaxes to a new equilibrium. The speed of relaxation is measured by a physical method such as spectroscopy, and analysis of the results leads to the reaction rate.

The most common way of changing the external conditions is to change the temperature, and the method is called the temperature-jump, or T-jump, method. Techniques have been developed for raising the temperature of a tiny reaction vessel by a few degrees in less than 100 ns. The method is therefore not suitable for the fastest processes, which can be studied by flash photolysis, but many purely chemical processes are suitable for the T-jump technique, which has provided valuable kinetic information.

Other experimental techniques are used for the study of rapid processes. Ultrasonic methods have been used for processes occurring with half-lives in the microsecond (μs; 1 μs = 10^{-6}second) and nanosecond ranges. Nuclear magnetic resonance has also been used for certain types of reactions.

Some Kinetic Principles

The kinetic behaviour of an ordinary chemical reaction is conventionally studied in the first instance by determining how the reaction rate is influenced by certain external factors such as the concentrations of the reacting substances, the temperature, and sometimes the pressure. For a reaction in which two substances A and B react with each other, it is sometimes found that the reaction rate is proportional to the concentration of A, represented by [A], and to the concentration of B, or [B]. In that case the reaction is said to be a second-order reaction; it is first order in [A] and first order in [B]. In such a case the reaction rate v can be expressed as $v=k[A][B]$, where k is a constant, known as the rate constant for the reaction.

This is just one of many types of kinetics that can be observed. A substance A that changes into another substance may obey a kinetic equation of the form v = k[A], which is a first-order reaction. It is important to recognize that the kinetics of a reaction does not always correspond in a simple way to the balanced chemical equation for the reaction. Thus, if a reaction is of the form $A+B \rightleftharpoons Y+Z$, the reaction is not necessarily second-order in both directions. This is in contrast to the situation with the equilibrium constant for the reaction, which corresponds to the balanced equation. The reason why the kinetic law is different is that the reactions in the forward and reverse directions may occur by stepwise mechanisms that lead to a different and usually more complex kinetic equation.

Sometimes reaction rates depend on reactant concentrations in a more complicated way. This is a clear indication that a reaction happens in several steps.

The effect of temperature on reaction rates provides much information about reaction mechanisms. Understanding of this effect owes much to the ideas of the Dutch physical chemist Jacobus Henricus van't Hoff and the Swedish chemist Svante August Arrhenius. Their equation for the dependence of a rate constant k on the absolute temperature T is $k=A\,exp\,(-E/RT)$, where R is the molar gas constant and A and E are quantities that are different for each reaction. This equation has come to be called the Arrhenius equation, although, as Arrhenius acknowledged when he applied it in 1889, it was first suggested by van 't Hoff in 1884. According to this relationship, a plot of the logarithm of the rate constant against the reciprocal of the absolute temperature should yield a straight line. From the slope and intercepts of the line, it is possible to calculate the value of the kinetic parameters A and E. The Arrhenius relationship applies satisfactorily to most reactions and indeed to many physical processes; however, various complications may cause it to fail.

If the reaction between two molecules is an elementary one, occurring in a single step, a simple interpretation of the Arrhenius equation can be given. The quantity A is related to the frequency of collisions between the reacting molecules. The quantity E, known as the activation energy for the reaction, results from the fact that there is an energy barrier to reaction. If E was zero, k would be equal to A, which means that the reaction would occur every time a collision occurred between the reactant molecules. This is the case for reactions in which no chemical bond is broken, such as the combination of atoms.

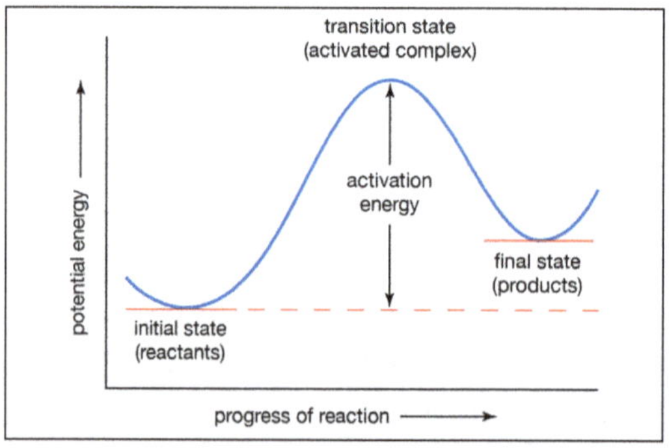

In figure the Arrhenius equation, the activation energy (E) represents the minimum amount of energy required to transform reactants into products in a chemical reaction. On a potential energy curve, the value of the activation energy is equivalent to the difference in potential energy between particles in an intermediate configuration (known as the activated complex, or transition state) and particles of reactants in their initial state. The activation energy thus can be visualized as a barrier that must be overcome by reactants before products can be formed.

For reactions in which a chemical bond is broken, on the other hand, the activation energy E is not zero but has a value that is often a tenth or so of the energy required to break the bond. A simple and essentially correct explanation of the activation energy was suggested by Arrhenius, who pointed out that, for many reactions, raising the temperature by 10 °C (18 °F) doubles the reaction rate. This increase cannot be caused by the increase in the frequency of collisions between colliding molecules, since the frequency does not increase sufficiently with a rise in temperature. Arrhenius suggested that when reactants A and B react together, they first form a highly energized intermediate that is denoted as $AB^{\ddagger}$, which subsequently gives the products of reaction:

$$A + B \rightleftharpoons AB^{\ddagger} \rightarrow Y + Z.$$

If the intermediate complex (also called the activated complex) $AB^{\ddagger}$ is of high energy, it is formed only in small amounts. According to the Boltzmann principle, the fraction of molecules having energy greater than E is exp (−E/RT), which provides an explanation of the appearance of this fraction in the Arrhenius equation. The interpretation of the equation is thus that only those molecules having energy greater than E are able to undergo reaction; other collisions are ineffective, and the reactant molecules merely separate unchanged.

Composite Reaction Mechanisms

Various lines of evidence are used to determine if a reaction occurs in more than one step. Suppose that the kinetic equation for the reaction does not correspond to the balanced equation for the reaction. A simple example is the reaction between hydrogen and iodinechloride, with the formation of iodine and hydrogen chloride:

$$H_2 + 2ICl \rightarrow I_2 + 2HCl$$

To make the equation balance, the reaction must be written as shown, with two iodine chloride molecules reacting with a single hydrogen molecule. If this reaction occurred in a single elementary step, the rate would be proportional to the first power of the hydrogen concentration and the square of the iodine chloride concentration. Instead, however, the rate is found to be proportional to both concentrations to the first power, so that it is a second-order reaction:

$$v = k[H_2][ICl].$$

This can be explained if there is initially a slow reaction between one hydrogen molecule and one of iodine chloride $H_2 + ICl \rightarrow HI + HCl(\text{slow})$ followed by a rapid reaction between the hydrogen iodide formed and an additional molecule of iodine chloride:

$$HI + ICl \rightarrow HCl + I_2(\text{fast}).$$

If the second reaction is fast, the hydrogen iodide is removed as fast as it is formed. The rate of the second reaction therefore has no effect on the overall rate, which is the rate of the first step. This mechanism therefore explains the kinetic behaviour but does not prove it; other, more complicated schemes could be devised, but, until there is further evidence, it is expedient to accept the simple mechanism. This is an example of a consecutive reaction, which occurs in two steps, with the intermediate playing a role.

Another piece of evidence for a composite mechanism is the detection of reaction intermediates. In such a case, a reaction scheme must be devised that will account for these intermediates. Sometimes an intermediate can be a fairly stable substance. In other cases the intermediates are unstable species such as atoms and free radicals (fragments of molecules) that subsequently undergo rapid reactions. Free radicals can be detected by spectroscopy and other means. When organic molecules are raised to high temperatures, they decompose into smaller molecules, and organic free radicals have often been detected as intermediates. In an explosion, such as that between hydrogen and oxygen, free radicals such as hydroxyl can be detected.

Composite reaction mechanisms are of various kinds. Aside from the simple consecutive schemes, there are some special mechanisms that give rise to oscillatory behaviour: the amount of a product continuously rises and falls over a period of time. The conditions for this behaviour are that there must be at least two species involved in the reaction and there must be feedback, which means that products of the reaction affect the rate. There are also reaction mechanisms that give rise to what is technically known as chaos, or catastrophe. With such reactions it is impossible to predict the outcome. Chaotic conditions also require that there be feedback and that at least three species be involved.

Sometimes a complex reaction mechanism involves a cycle of reactions such that certain intermediates consumed in one step are regenerated in another. For example, the accepted mechanism of the reaction between hydrogen and bromine, which can be written as $H_2 + Br_2 \rightarrow 2HBr$, includes the steps:

$$Br + H_2 \rightarrow HBr + H$$
$$H + Br_2 \rightarrow HBr + Br.$$

In the first of these steps a bromine atom is consumed, but in the second a bromine atom is regenerated. This pair of reactions can thus occur with the production of two molecules of hydrogen bromide, the product of the reaction, without loss of bromine atoms. This pair of reactions is called a cycle of reactions, and it can occur a number of times, in which case the reaction is referred to as a chain reaction. The two reactions in which bromine is regenerated are known as the chain-propagating steps. The average number of times the pair of steps is repeated is known as the chain length.

One necessary condition for a proposed reaction mechanism to be correct is that it must account for the overall kinetic behaviour of the reaction—in particular, for the dependence of the reaction rate on the reactant concentrations. For any proposed reaction mechanism, it is possible to write down equations for the rate of each step in terms of the reactant concentration and then to solve the equations for the overall rate. A practical difficulty arises, since no exact mathematical solution is possible for all except the simplest of mechanisms. If one has values for the rate constants, solutions can be obtained with a computer, but explicit rate equations provide more insight into the reactions. One therefore looks for approximate solutions of the equations. One of these is provided by the steady-state treatment, which is applicable if (and only if) the intermediates are species that can be present only at low concentrations. If this condition is satisfied by an intermediate, the rate of change of its concentration during the course of reaction is always small and, as a good approximation, can be assumed to be zero, which means that the intermediate exists in a steady state. This approximation may safely be applied to atoms and free radicals present as reaction intermediates. With this approximation it is usually possible to obtain a reliable approximate equation for the overall reaction rate in terms of reactant concentrations. If this agrees with the experimental behaviour, the mechanism is accepted.

One situation to which the steady-state treatment does not apply is when a reaction is an explosion. Explosions occur because the concentration of intermediates does not remain steady during the course of reaction but rises to a high value so that the reaction goes out of control. This occurs if the reaction mechanism involves a special kind of chain called a branching chain. In the hydrogen-oxygen explosion, for example, the following reaction is known to occur:

$$H + O_2 \rightarrow OH + O.$$

In this step a single chain carrier hydrogen atom has produced two chain carriers: a hydroxyl group and an oxygen atom. The number of chain carriers increases rapidly and leads to an explosion.

Theories of Reaction Rates

Two different theoretical approaches to chemical kinetics have led to an understanding of the details of how elementary chemical reactions occur. Both of these are based on the idea of potential-energy surfaces, which are models showing how the potential energy of a reaction system varies with certain critical interatomic distances. The course of an elementary reaction is represented by the movement of the system over the potential-energy surface. One theoretical approach to the problem involves studying the region of the potential-energy surface that corresponds to the highest point on the energy barrier that separates the reactants from the products. This approach is relatively simple and leads to explicit general expressions for the reaction rate. The second approach involves considering the dynamics of the motion of the system over the potential-energy surface.

Transition-state Theory

The idea of a potential-energy surface sprang from the ideas of Dutch physical chemist Jacobus Henricus van't Hoff and Swedish physicist Svante August Arrhenius that were put forward to explain the effect of temperature on reaction rates. An important advance was made in 1931 by American chemist Henry Eyring and British chemist Michael Polanyi, who constructed, on the basis of quantum mechanics, a potential-energy surface for the simple reaction:

$$H^{\alpha} + H^{\beta}-H^{\gamma} \rightarrow H^{\alpha}-H^{\beta}-H^{\gamma} \rightarrow H^{\alpha}-H^{\beta} + H^{\gamma}$$

For convenience the labels α, β, and γ are added as superscripts. When this reaction occurs, an atom H^{α} attacks a hydrogen molecule $H^{\beta}-H^{\gamma}$ and abstracts one of the hydrogen atoms from it. As the bond begins to form, the $H^{\beta}-H^{\gamma}$ bond becomes more and more extended and finally breaks. Somewhere along the reaction path, there is a particular intermediate state corresponding to the maximum value of the potential energy.

This particular intermediate state is usually designated by the superscript. It is known as an activated complex and plays an important role in what has come to be called transition-state theory, developed independently in 1935 by Eyring, Polanyi, and English physical chemist M.G. Evans. The essential feature of the theory is that the activated complexes are considered to be formed from the reactants in a state in which they are in equilibrium with the reactants. Thus, the above reaction can be written as:

$$H^{\alpha} + H^{\beta}-H^{\gamma} \rightarrow H^{\alpha}-H^{\beta}-H^{\gamma\ddagger} \text{ (at equilibrium with } H^{\alpha} + H^{\beta}-H^{\gamma}) \rightarrow H^{\alpha}-H^{\beta} + H^{\gamma}$$

Since the activated complexes are in equilibrium, their concentration can be expressed in terms of the concentrations of the reactants. The reaction rate is this concentration multiplied by the frequency with which they form products, which is known from kinetic theory. Despite the approximations involved in transition-state theory, it has been successful in providing an insight into how chemical reactions occur and how their rates depend on various factors.

Molecular Dynamics

The second theoretical approach to chemical kinetics is referred to as molecular dynamics, or reaction dynamics. It is a more detailed treatment of reactions and is designed to investigate the atomic motions that occur during a chemical reaction and the quantum states of the reactant and product molecules. Such studies are important in testing the validity of transition-state theory and similar treatments. Also, there are important practical applications of kinetics, such as reactions occurring in lasers, for which information about the energy states of the products of a chemical reaction is needed; this information is not provided by transition-state theory but is an important outcome of molecular dynamics.

Consider a simple reaction of the type A + B−C → Y + Z, where A is an atom and B−C is a diatomic molecule. A dynamical calculation would first involve calculating, using quantum mechanics, a potential-energy surface that gives the potential energy corresponding to a set of initial configurations. One can make dynamical calculations for a variety of vibrational states of the reactant B−C and for a variety of translational energies. Ideally the calculations would be based on quantum mechanics, but this proves difficult, and often classical mechanics is used. Except for certain types of reactions where quantum effects are important, it appears that for many reaction systems the error involved in neglecting quantum effects is not great. In principle the dynamical calculations should give more-reliable results than any other treatment; however, for any but the simplest reactions, the computer calculations are time-consuming, and, since many approximations must be made to save computer time, there is often some uncertainty about the results.

Much work has been done along these lines, and on the whole the agreement with both experiment and transition-state theory is satisfactory. In addition, some calculations suggest important generalizations about reactions. For example, the form of the potential-energy surface greatly influences whether energy released in a reaction resides in the vibrations of the product molecules or in their kinetic energy of translation. Experimental studies of simple reactions, particularly some that involve the study of light emission in reactions (chemiluminescence) and that use narrow molecular beams, have also contributed to knowledge of chemical reactions.

References

- Physical-Chemistry: chemistryexplained.com, Retrieved 2 May, 2019
- Branches-of-physical-chemistry: azchemistry.com, Retrieved 12 March, 2019
- Henrys-law: lumenlearning.com, Retrieved 14 June, 2019
- Henrys-law, chemistry: byjus.com, Retrieved 16 January, 2019
- Definition-of-raoults-law: thoughtco.com, Retrieved 19 July, 2019
- Changes-In-Vapor-Pressure, Raoult's-Law, Ideal-Solutions, Solutions-and-Mixtures, Physical-Properties-of-Matter, Physical-and-Theoretical-Chemistry: libretexts.org, Retrieved 8 February, 2019
- Chemical-kinetics, science: britannica.com, Retrieved 18 April, 2019

2

Ideal and Real Gases

The hypothetical gases whose molecules do not interact with each other and occupy negligible space are known as ideal gases. The real gases have interactions between their molecules and thus do not behave like ideal gases. Some of the laws which govern gases are Boyle's law and Charles law. This chapter has been carefully written to provide an easy understanding of the varied facets of ideal and real gases.

Ideal Gas

A hypothetical gas which obeys Boyle's law exactly at all temperatures and pressures, and which has internal energy that depends only upon the temperature. Measurements upon real gases are extrapolated to zero pressure to obtain results in agreement with theories relating to an ideal gas, esp in thermometry.

An ideal gas is a gas whose pressure *P*, volume *V*, and temperature *T* are related by the ideal gas law:

$$PV = nRT$$

where, *n* is the number of moles of the gas and *R* is the ideal gas constant.

An ideal gas is also known as a perfect gas. Ideal gases are defined as having molecules of negligible size with an average molar kinetic energy dependent only on temperature. At a low temperature, most gases behave enough like ideal gases that the ideal gas law can be applied to them.

Types of Ideal Gas

Range of validity of classical ideal gas:

For a classical ideal gas, we derived the partition function:

$$Z = \frac{Z_1^N}{N!}, \quad Z1 = \frac{V}{\lambda_{th}^3} = V\left(\frac{2\pi m k_B T}{h^2}\right)^{3/2},$$

Where, the length scale $\lambda_{th} \equiv \frac{h}{2\pi m k_B T}$ is determined by the particle mass and the temperature.

When does this break down?

1. If 'idealness' fails, i.e. if interactions become important. This is interesting and important but out of bounds for 8.044. We'll still assume non-interacting particles.

2. If 'classicalness' fails. Even with no interactions, at low enough temperatures, or high enough densities, quantum effects modify Z.

An argument that classicalness must fail comes by thinking harder about λ_{th}, the "thermal de Broglie wavelength". For a classical ideal gas, we know the RMS momentum from equipartition.

$$\left\langle \frac{p^{-2}}{2m} \right\rangle = \frac{3}{2} k_B T \Rightarrow p_{RMS} \equiv \sqrt{\langle p^{-2} \rangle} = \sqrt{3mk_B T}\ .$$

We could have defined:

$$\tilde{\lambda}_{th} \equiv \frac{h}{p_{RMS}} = \frac{h}{\sqrt{3mk_B T}}$$

There's no significance to the numerical prefactor of $\frac{1}{\sqrt{2\pi}}$ instead of $\frac{1}{\sqrt{3}}$.

Now recall the significance of λ_{dB} in QM:

$$\lambda_{dB} = \frac{h}{p} \sim \text{ minimum size of a wavepacket with momentum p.}$$

So, we can infer that:

$\lambda_{th} \sim$ minimum size of quantum wavepackets describing atoms in a quantum ideal gas.

The classical picture of atoms as billiard balls with well-defined trajectories only makes sense if:

$$\lambda_{th} \ll \text{typical spacing between particles} = \left(\frac{V}{N}\right)^{1/3}$$

And this inequality $\lambda_{th} \sim \frac{1}{\sqrt{T}} \ll \left(\frac{V}{N}\right)^{1/3}$ is violated at low T or high density. The above inequality is the condition for validity of the classical treatment.

System	T	$\left(\frac{V}{N}\right)^{1/3} / \lambda_{th}$	Conclusion
Air	300K	180	Classical
Liquid ^{4}He	4K	0.86	Not classical
Conduction e^- in Cu	300K	0.053	Not classical

A second derivation of the same criterion, this time in momentum space:

Consider solutions to the Schrödinger equation in an $L \times L \times L$ box. The energy eigenstates of one particle are:

$$\varphi_{\vec{m}}(\vec{x}) = \sin(k_x x) \sin(k_y y)\ \sin(k_z z) \text{ with } \begin{matrix} k_x \\ y \\ z \end{matrix} = \frac{\pi}{L} \begin{matrix} m_x, \\ y \\ z \end{matrix} \quad \text{and } \begin{matrix} m_x \\ y \\ z \end{matrix} = 1, 2, 3\ldots$$

with energy eigenvalue:

$$\in = \frac{\hbar^2}{2m}\ (k_x^2 + k_y^2 + k_z^2).$$

Let's ask under what circumstances we can ignore the QM.

The mean energy per particle in a classical ideal gas is $\frac{3}{2}k_BT$. How many 1-particle states with energies like this are there to put our N particles into? The number of states with $\in \le \frac{3}{2}k_BT$ is:

$$N(\in) = \frac{\frac{1}{8}\left(\frac{4}{3}\pi\left(\frac{2m\frac{3}{2}k_BT}{\hbar^2}\right)^{3/2}\right)}{(\pi/L)^3} = \frac{\text{volume of octant of sphere in } \vec{k}-\text{space with radius} \left(\frac{2mE}{\hbar^2}\right)^{1/2}}{\text{volume in } \vec{k}-s\,\text{pace per allowed grid point}}$$

$$= V\left(\sqrt{\frac{mk_BT}{h}}\right)^3 \times 4\sqrt{3}\pi = \frac{V}{\lambda_{th}^3} \times \frac{\sqrt{3\cdot 2}}{\sqrt{\pi}}$$

So, when $N \gg \frac{V}{\lambda_{th}^3}$, the number of particles is big compared to the number of possible oneparticle states for the particles to sit in. So we need to worry about things like the Pauli Exclusion principle, which prevents us from putting more than one particle in each state. $N \ll \frac{V}{\lambda_{th}^3}$, the classical analysis is fine – we needn't worry about the particles needing to occupy the same 1-particle state.

Quantum Systems with Many Indistinguishable Particles

Consider two particles: Their state can be described by a wave function $\psi(x_1, x_2)$. (We won't worry right now about how many coordinates of each particle (e.g. how many dimensions) we have to specify.) If the particles are indistinguishable, then,

$$|\psi(x_1, x_2)|^2 = \text{Prob}(\text{finding a particle at } x_1 \text{ and a particle at } x_2).$$

article is where. Indistinguishability requires:

$$|\psi(x_1, x_2)|^2 = |\psi(x_2, x_1)|^2.$$

Swapping the arguments twice should give back the same ψ, not just the same $|\psi|^2$:

$$\psi(x_1, x_2) = \pm\psi(x_2, x_1).$$

Bosons

- Particles for which $\psi(x_1, x_2) = +\psi(x_1, x_2)$ i.e., the wave function is symmetric.
- It is a fact (observed experimentally, understood via quantum field theory) that they have integer spin. e.g., hydrogen atoms, ^{4}He, photons, phonons, magnons, gluons, Higgs bosons.
- The stat mech of a gas of them was developed by Bose and Einstein, so in the context of stat mech, these are called "Bose-Einstein statistics".

Fermions

- Particles for which $\psi(x_1, x_2) = -\psi(x_1, x_2)$ i.e., the wave function is anti-symmetric.
- It is a fact (observed experimentally, understood via quantum field theory) that they have half-integer spin (1/2, 3/2...). E.g.: electron, proton, neutron, ^{3}He, ^{7}Li.
- The stat mech of a gas of them was developed by Fermi and Dirac, hence "Fermi-Dirac statistics."

For non-interacting Bose or Fermi particles: energy eigenstates are always (symmetric or antisymmetric) linear combinations of products of single-particle energy eigenstates.

Wavefunctions of Several Bosons or Fermions

Consider for example two indistinguishable quantum particles in a box. Label the possible states of one quantum particle in the box by a fancy label α which is a shorthand for all its quantum numbers, e.g., the wavenumbers (m_x, m_y, m_z) of its wavefunction.

If they are bosons, the wavefunction must be of the form:

$$\psi(x_1, x_2) = \underbrace{\underbrace{\phi_\alpha(x_1)}_{\substack{\text{an energy eigenstate of} \\ \text{one particle}}} \cdot \underbrace{\phi_\beta(x_2)}_{\substack{\text{another energy eigenstate} \\ \text{for one particle}}}}_{\substack{\text{not sym or antisym} \\ \text{hence not allowed}}} \underbrace{+}_{\substack{\text{makes it symmetric} \\ \text{OK state of bosons}}} \phi_\beta(x_1)\phi_\alpha(x_2).$$

For fermions:

$$\psi(x_1, x_2) = \phi_\alpha(x_1)\phi_\beta(x_2) \underbrace{-}_{\substack{\text{makes it antisymmetric} \\ \text{OK state of fermions}}} \phi_\beta(x_1)\phi_\alpha(x_2).$$

Working our way up to 3 indistinguishable particles:

$$\begin{aligned}\Psi(x_1, x_2, x_3) = {} & \phi_\alpha(x_1)\phi_\beta(x_2)\phi_\gamma(x_3) \pm \phi_\alpha(x_1)\phi_\gamma(x_2)\phi_\beta(x_3) \\ & \phi_\beta(x_1)\phi_\gamma(x_2)\phi_\alpha(x_3) \pm \phi_\beta(x_1)\phi_\alpha(x_2)\phi_\gamma(x_3) \\ & \phi_\gamma(x_1)\phi_\alpha(x_2)\phi_\beta(x_3) \pm \phi_\gamma(x_1)\phi_\beta(x_2)\phi_\gamma(x_3)\end{aligned}$$

with + for bosons and – for fermions.

N indistinguishable Particles

1. Pick N single particle states α, β, γ....
2. There is exactly one symmetric combination. This is an energy eigenstate for N bosons.

3. If $\alpha, \beta, \gamma...$ are all different then there is exactly one antisymmetric combination. This is an energy eigenstate for N fermions.
   - This fact that "Fermions must all be in different single particle eigenstates" is the Pauli Exclusion Principle.
   - Many bosons can be in the same single-particle energy eigenstate.
   - If the particles were distinguishable, there would be N states for each choice of $\alpha, \beta, \gamma...$
   - If the particles are indistinguishable, there is (at most) one such state.

Occupation Number Representation of the Many-particle State

For either bosons or fermions, the state $\Psi(x_1, x_2...x_N)$ is fully specified by indicating:

- Which 1-particle states are occupied?
- If bosons, how many particles are in each 1-particle state? (For fermions, this number can only be 0 or 1).

Label the 1-particle states α (e.g. m_x, m_y, m_z for ideal gas, or n, ℓ, m for Hydrogen atoms). So: the state Ψ is specified by a set of integers called occupation numbers:

$$n_\alpha(\Psi) \equiv \#\text{ of particles in 1-particle state } \alpha$$

when the many-particle state is Ψ.

$$\text{Fermions: } n_\alpha \in \{0, 1\} \quad \text{Bosons: } n_\alpha \in \{0, 1, 2, 3...\}.$$

These numbers also specify N, E, ... , as follows:

$$\text{total } \# \text{ of particles in state } \Psi: N = \sum_\alpha n_\alpha(\Psi)$$

The sum here is over all possible states in which one of the particles could be; many of them will be unoccupied. This is not a sum over particles.

Similarly, if:

$$\in_\alpha = \text{the energy eigenvalue of the 1-particle energy eigenstate } \alpha.$$

Then the total energy in the many-particle state Ψ is:

$$\text{total energy in state } \Psi: E(\Psi) = \sum_\alpha n_\alpha(\Psi) \in_\alpha .$$

Statistical Mechanics of *N* non-interacting Bosons or Fermions

We'll use the canonical ensemble: an ensemble of copies of the system, all with the same N, T (Hence, E varies amongst the copies in the ensemble), in contact with a heat bath at temperature T.

$$\langle E\rangle = \sum_{i\in \text{ensemble}} \underbrace{E_i}_{\substack{\text{total energy of}\\ \text{N particles}}} \cdot \underbrace{p(\Psi_i)}_{\substack{\text{prob of finding the N-particle state}\\ \Psi_i \text{ in the ensemble}}}$$

$$= \sum_i \left(\sum_\alpha n_\alpha(\Psi_i) \epsilon_\alpha \right) \cdot p(\Psi_i)$$

$$= \sum_\alpha \epsilon_\alpha \cdot \underbrace{\left(\sum_i n_\alpha(\Psi_i) p(\Psi_i) \right)}_{\substack{\equiv \langle n_\alpha \rangle \\ \text{mean value of the occupation number} \\ \text{of the 1-particle state a in} \\ \text{the ensemble of N-particle states}}}$$

So, if we know $\langle n_\alpha\rangle \equiv \sum_i n_\alpha(\Psi_i)p(\Psi_i)$, then we know:

$$\langle E\rangle = \sum_\alpha \epsilon_\alpha \langle n_\alpha\rangle$$

A check on this formalism:

$$\sum_\alpha \langle n_\alpha\rangle = \sum_\alpha\sum_i n_\alpha(\Psi_i)p(\Psi_i) = \sum_i \underbrace{\left(\sum_\alpha n_\alpha(\Psi_i)\right)}_{=N} p(\Psi_i) = N\sum_i p(\Psi_i) = N$$

So, how do we calculate the average occupation numbers $\langle n_\alpha\rangle$?

$\langle n_\alpha\rangle$ for fermions:

$$\langle n_\alpha\rangle = \sum_{\text{all N-particle states } \Psi_i} n_\alpha(\Psi_i)p(\Psi_i) \qquad \left(p(\Psi) = Z(N)^{-1}\, e^{-\beta E(\Psi)}\right)$$

$$= \frac{1}{Z(N)}\sum_i n_\alpha(\Psi_i)e^{-\beta E_i}$$

$$= \frac{1}{Z(N)} \underbrace{\sum_{n_1=0}^{1}\sum_{n_2=0}^{1}\sum_{n_3=0}^{1}\cdots}_{\text{subject to the constraint } n_1+n_2+\ldots=N} n_\alpha\, e^{-\beta(n_1\epsilon_1+n_2\epsilon_2+\ldots)}$$

$$= \frac{1}{Z(N)} \underbrace{\sum_{n_1=0}^{1}\sum_{n_2=0}^{1}\sum_{n_3=0}^{1}\cdots}_{\text{subject to the constraint } n_1+n_2+\ldots=N} n_\alpha\, e^{-\beta(n_1\epsilon_1+n_2\epsilon_2+\ldots)}$$

We just sum over all occupation numbers, consistent with the fact that there are N particles altogether. One of these sums is special, so let's write the two terms $n_\alpha = 0$ and $n_\alpha = 1$ explicitly:

$$\langle n_\alpha\rangle = \underbrace{0}_{n_\alpha=0} \frac{1}{Z(N)} e^{-\beta\epsilon_\alpha} \underbrace{\sum_{n_1=0}^{1}\sum_{n_2=0}^{1}\sum_{n_3=0}^{1}\cdots}_{\text{doesn't include } n_\alpha \text{ which is 1 so the rest are subject to the constraint } n_1+n_2+\ldots=N-1} \underbrace{e^{-\beta(n_1\epsilon_1+n_2\epsilon_2+\ldots)}}_{\text{excluding } n_\alpha\epsilon_\alpha}$$

Repackage in a sneaky way which takes advantage of $n_\alpha = 0$ or 1:

$$\langle n_\alpha \rangle = \frac{1}{Z(N)} e^{-\beta \epsilon_\alpha} \underbrace{\sum_{n_1=0}^{1} \sum_{n_2=0}^{1} \sum_{n_3=0}^{1} \cdots}_{\substack{\text{includes } n_\alpha \text{ subject to the} \\ \text{constraint } n_1 + n_2 + \ldots = N-1}} (1 - n_\alpha) \underbrace{e^{-\beta(n_1 \epsilon_1 + n_2 \epsilon_2 + \ldots)}}_{\text{includes } n_\alpha \epsilon_\alpha}$$

Notice that the $n_\alpha = 1$ term in the summation gives zero; this means it doesn't matter if we include it in the Boltzmann factor. The $n_\alpha = 0$ term in the summation gives what we had before.

$$\begin{aligned}\langle n_\alpha \rangle &= \frac{1}{Z(N)} e^{-\beta \epsilon_\alpha} \left(Z(N-1) - \langle n_\alpha \rangle_{\text{for } N-1 \text{ particles}} Z(N-1) \right) \\ &= \frac{Z(N-1)}{Z(N)} e^{-\beta \epsilon_\alpha} \left(1 - \langle n_\alpha \rangle_{\text{for } N-1 \text{ particles}}\right)\end{aligned}$$

The Fermi-Dirac Distribution

We want to rewrite our expression for $\langle n_\alpha \rangle_{\text{fermions}}$ as a function of T, μ.

To do this, recall that:

$$\mu \equiv F(T, V, N) - F(T, V, N-1) = -k_B T \; (\ln Z(N) - \ln Z(N-1))$$

$$\Rightarrow \frac{Z(N-1)}{Z(N)} = e^{\mu / k_B T}$$

Now, in a box with three particles, $\langle n_\alpha \rangle_{N=3} \neq \langle n_\alpha \rangle_{N-1=2}$. But for $N \sim 10^{25}$,

$$\langle n_\alpha \rangle_N \approx \langle n_\alpha \rangle_{N-1} \quad \text{for } N \gg\gg 1.$$

$$\Rightarrow \langle n_\alpha \rangle = e^{\mu / k_B T} \; e^{-\epsilon_\alpha / k_B T} \cdot \left(1 - \langle n_\alpha \rangle\right)$$

Solve this for $\langle n_\alpha \rangle : \langle n_\alpha \rangle \left(1 + e^x\right) = e^x$, with $x \equiv \beta \; (\mu - \epsilon_\alpha)$

$$\Rightarrow \langle n_\alpha \rangle = \frac{e^x}{1 + e^x} = \frac{1}{e^{\beta(\epsilon_\alpha - \mu)} + 1} \text{ for fermions.}$$

What's μ? It's determined from the relation between μ and N:

$$N = \sum_\alpha \langle n_\alpha \rangle = \sum_\alpha \frac{1}{e^{\beta(\epsilon_\alpha - \mu)} + 1}$$

Notice: each $\langle n_\alpha \rangle \in [0, 1]$, since $e^{-x} \in [0, \infty]$ for real x. This is the Pauli Principle in action. The extremes are realized in the $T \to \infty, \; 0$ limits, if we hold μ fixed. More on this below:

$\langle n_\alpha \rangle$ for Bosons (the Bose-Einstein distribution)

$$\langle n_\alpha \rangle = \frac{1}{Z(N)} \underbrace{\sum_{n_1=0}^{\infty} \sum_{n_2=0}^{\infty} \sum_{n_3=0}^{\infty}}_{\text{with } n_1+n_2+\ldots=N} \cdots \; n_\alpha \; e^{-\beta(n_1 \epsilon_1 + n_2 \epsilon_2 + \ldots)}$$

$$\text{one special sum}: \sum_{n_\alpha=0}^{\infty} n_\alpha e^{-\beta n_\alpha \epsilon_\alpha}$$

No contribution from $n_\alpha = 0$. Tricky step : let $n_\alpha^{'} \equiv n_\alpha - 1$

$$= \frac{1}{Z(N)} \underbrace{\sum_{n_1=0}^{\infty} \sum_{n_2=0}^{\infty} \sum_{n_3=0}^{\infty}}_{\text{with } n_1+n_2+\ldots+n_\alpha^{'}+\ldots=N-1} \cdots \quad (n_\alpha^{'}+1)\, e^{-\beta \epsilon_\alpha (1+n_\alpha^{'}) - \beta(n_1 \epsilon_1 + \ldots)}$$

$N - 1$ in all states including ϕ_α, at least one particle in ϕ_α.

$$= \frac{e^{-\beta \epsilon \alpha}}{Z(N)} \left(Z(N-1) + \langle n_\alpha \rangle_{\text{for N - 1 particles}} \cdot Z(N-1) \right)$$

The only difference from fermions is the plus sign. Making the same steps following equation:

$$\langle n_\alpha \rangle = \frac{1}{Z(N)} e^{-\beta \epsilon_\alpha} \left(Z(N-1) - \langle n_\alpha \rangle_{\text{for } N-1 \text{ particles}} Z(N-1) \right)$$

$$= \frac{Z(N-1)}{Z(N)} e^{-\beta \epsilon_\alpha} \left(1 - \langle n_\alpha \rangle_{\text{for } N-1 \text{ particles}}\right)$$

$$\Rightarrow \langle n_\alpha \rangle = \frac{1}{e^{(\epsilon_\alpha - \mu)/k_B T} - 1} \text{ for Bosons.}$$

The minus sign in the denominator is very significant: with this sign, the denominator can become small, and so $\langle n_\alpha \rangle_{\text{Bosons}}$ can be larger than unity.

Again μ is determined from the relation between μ and N:

$$\text{N} = \sum_\alpha \langle n_\alpha \rangle = \sum_\alpha \frac{1}{e^{\beta(\epsilon_\alpha - \mu)} - 1}$$

Classical Ideal Gas Limit of Quantum Ideal Gas

If $e^{(\epsilon_\alpha - \mu)/k_B T} \gg 1$ then:

$$\langle n_\alpha \rangle_{fermions} = \langle n_\alpha \rangle_{bosons} \approx e^{-(\epsilon_\alpha - \mu)/k_B T} \equiv \langle n_\alpha \rangle_{Boltzmann} \ll 1$$

This factor gives the occupation numbers for "Boltzmann statistics".

Claim: this is the classical limit. A weak check on this claim is the fact that the B and F distributions reduce to the same expression in this limit. To show: $\langle n_\alpha \rangle_{Boltzmann}$ are the classical occupation numbers in thermal equilibrium.

1. Classical when $e^{-\mu/k_BT} \gg 1 \Rightarrow e^{+\mu/k_BT} \ll 1$. A classical ideal gas has:

$$-\frac{\mu}{k_BT} = ln\left(\frac{V}{N}\frac{1}{\lambda_{th}^3}\right)$$

So,

$$e^{-\mu/k_BT} \ll 1 \Rightarrow \frac{V}{N}\frac{1}{\lambda_{th}^3} \gg 1,$$

which is our classicality criterion from before.

2. Now that we know we're in the right limit, let's examine the Boltzmann occupation number, and show that it's the classical expectation.

$$\langle n_\alpha \rangle_{Boltzmann} = e^{(\mu-\epsilon_\alpha)/k_BT}$$

Previously we wrote everything in terms of N, so let's do that now:

$$N = \sum_\alpha \langle n_\alpha \rangle = e^{\mu/k_BT} \underbrace{\sum_\alpha e^{-\beta\epsilon_\alpha}}_{=Z_1}$$

$$\Rightarrow e^{\mu/k_BT} = \frac{N}{Z_1} \text{ for Boltzmann statistics}$$

$$\mu = -k_BT \; ln\,\frac{Z_1}{N}$$

This is the familiar classical ideal gas result, if we can show that Z_1 is indeed the classical ideal gas 1-particle partition function. We also have:

$$\langle n_\alpha \rangle_{Boltzmann} = \frac{N}{Z_1} e^{-\epsilon_\alpha/k_BT},$$

which is a familiar fact from the canonical ensemble. That is, it is equivalent to the happy statement:

$$\text{prob}\left(1 \text{ atom is in state } \alpha\right) = p(\alpha) = \frac{e^{-\beta\epsilon_\alpha}}{Z_1} = \frac{\langle n_\alpha \rangle_{Boltzmann}}{N}.$$

Also, from the definition of chemical potential,

$$Z(N) = e^{-\mu/k_BT} Z(N-1).$$

For the particular case of Boltzmann statistics:

$$Z(N) = \frac{Z_1}{N} Z(N-1) \text{ this becomes a recursion relation!}$$

$$= \frac{Z_1}{N}\frac{Z_1}{N-1} Z(N-2) = \frac{Z_1}{N}\frac{Z_1}{N-1}\frac{Z_1}{N-2} Z(N-3) = \ldots$$

$$Z(N) = \frac{(Z_1)^N}{N!}$$

We have now derived this result from the classical limit of the quantum result.

Finally, we evaluate Z_1:

$$Z_1 = \sum_\alpha e^{-\beta \epsilon_\alpha} \text{ where } \alpha \text{ is a label on single-particle energy eigenstates}$$

$$= \sum_{k_x}\sum_{k_y}\sum_{k_z} e^{-\frac{\hbar^2}{2mk_BT}\left(k_x^2+k_y^2+k_z^2\right)}$$

$$= \left(\sum_{kx} e^{-\frac{\hbar^2}{2mk_BT}k_x^2}\right)^3 \qquad \text{with } k_x = m_x \frac{\pi}{L}, m_x = 1, 2, 3\ldots$$

In the high temperature or low-density limit, the steps in k_x are negligible and we can approximate the sum as an integral.

$$Z_1 = \left(\int_0^\infty dk_x \frac{L}{\pi} e^{-\frac{\hbar^2}{2mk_BT}k_x^2}\right)^3 = \left(\int_{-\infty}^\infty dk_x \frac{L}{2\pi} e^{-\frac{\hbar^2}{2mk_BT}k_x^2}\right)^3$$

$$= \left(\frac{L}{2\pi}\left(\frac{2\pi mk_BT}{\hbar^2}\right)^{1/2}\right)^3 = V\left(\frac{2\pi mk_BT}{\hbar^2}\right)^{3/2}$$

$$= \frac{V}{\lambda_{th}^3}.$$

Finally, this is the fulfillment of the promise that the factor of $\frac{1}{h^3}$ that we apparently arbitrarily used to normalize our classical phase space integrals would come from QM, i.e. it's what we get when we start with the correct QM answer and take the classical limit.

Ultra-high Temperature ($k_BT \gg mc^2$ and $k_BT \gg \mu$) limit

Let's evaluate the average energy $\langle E\rangle$ for an ultrarelativistic gas of fermions or bosons at high temperature, in particular at temperature bigger than any other energy scale: $kT \gg mc^2, kT \gg \mu$.

$$\langle E \rangle = \sum_{\alpha} \epsilon_{\alpha} \langle n_{\alpha} \rangle = \int_0^{\infty} \frac{\frac{1}{8} 4\pi k^2 dk}{(\pi / L)^3} \epsilon(k) \langle n_k \rangle$$

where,

$$\epsilon(k) = \sqrt{p^2c^2 + m^2c^4} = \sqrt{\hbar^2c^2k^2 + m^2c^4} \approx \hbar ck$$

This last approximation will be valid in the high-temperature regime.

and,

$$\langle n_k \rangle = \frac{1}{e^{(\epsilon(k)-\mu)/k_BT} \pm 1} \quad \begin{matrix} +: \text{fermions} \\ -: \text{bosons} \end{matrix}$$

So,

$$\frac{\langle E \rangle}{V} = \frac{\hbar c}{2\pi^2} \int_0^{\infty} \frac{dkk^3}{e^{(\hbar ck - \mu)/k_BT} \pm 1}.$$

For $k_BT \gg \mu$, we can ignore the contribution from μ. Then we can non-dimensionalize the integral by introducing $x \equiv \frac{\hbar ck}{k_BT}$, in terms of which:

$$\frac{\langle E \rangle}{V} = \frac{(k_BT)^4}{2\pi^2(\hbar c)^3} \underbrace{\int_0^{\infty} \frac{dxx^3}{e^x \pm 1}}_{= \begin{cases} \frac{7\pi^4}{15}, \text{ for fermions} \\ \frac{8\pi^4}{15}, \text{ for bosons} \end{cases}}$$

$$\Rightarrow \frac{\langle E \rangle}{V} = \frac{\pi^2}{30} \frac{(k_BT)^4}{(\hbar c)^3} \times \begin{cases} 1, \text{for bosons} \\ \frac{7}{8}, \text{ for fermions} \end{cases}$$

Observation: the blackbody radiation answer is obtained from this formula with two species of massless bosons with $\mu = 0$.

Reinterpretation: $\langle n \rangle$ for the harmonic oscillator modes of EM field ≡ the occupation number for $\vec{k}$ -space modes of these bosons.

- These bosons are called photons.
- There are two species because of the two polarizations.
- They are massless because the dispersion relation for EM waves is $\omega = ck$; there is no parameter with dimensions of mass in Maxwell's equations. This a deep fact. A mass term

in the energy density of the EM field would look something like $m^2 A_\mu A^\mu$; but this is not invariant under gauge transformations.

- Unlike the number of atoms in a box, the number of photons is not a conserved quantity. This forces the chemical potential to be zero – there's no need for a constraint fixing N, and hence no need for the Lagrange multiplier μ.

Relatedly, μ = 0 says there is no free energy cost to adding one more photon. The fact that they are massless means we can add photons with $\vec{k}$ = 0 at zero energy cost, since $\in(k=0)=0$. (The fact that they are bosons means we can keep doing this over and over, even if the mode already has nonzero occupation number).

As T grows, the total number of photons = $\sum_\alpha \langle n_\alpha \rangle$ also grows. This is distinct from the case of e.g. Helium atoms, where the chemical potential $\mu(T)$ must adjust to keep the number of atoms fixed.

Generally, heating something up by an extreme amount is a good way to figure out what are its constituents. Conveniently for particle physicists, the whole universe somehow got heated up quite a bit in the past. The early universe is a great source of examples of ultrarelativistic bosons and fermions at high temperature. For example, at $k_B T = 4\times10^8\ eV = 400 MeV\ (i.e.\ T \sim 10^{12}\cdot 5\ K)$, which was the situation at about 1 microsecond after the Big Bang, an accounting of the energy density begins by thinking about a gas of photons, electrons, positrons, quarks, antiquarks, gluons.

People have recently recreated similar conditions by colliding heavy ions at very high speeds. They have created a new state of matter called Quark-Gluon Plasma, which turns out to be a liquid at the temperatures and densities.

Fermions at Low Temperatures

Low temperatures is where quantum statistics – the differences between boson, fermions, and 'Boltzmons' – are most dramatic.

For fermions in thermal equilibrium, the occupation number of the single-particle state α is:

$$\langle n_\alpha \rangle = \frac{1}{e^{(\epsilon_\alpha-\mu)/k_B T}+1}.$$

The only property of the single-particle state that enters is its energy, so we may relabel this (as is conventional) as $f(\in) \equiv \frac{1}{e^{(\in-\mu)/k_B T}+1}$, the Fermi distribution function.

This one function contains the physics of metals, semiconductors, white dwarves, neutron stars.

Consider first the extreme $T \to 0$ limit of this function. It looks like this: → → → → →

A useful fact: $f(\in = \mu) = \frac{1}{2}$ at any temperature.

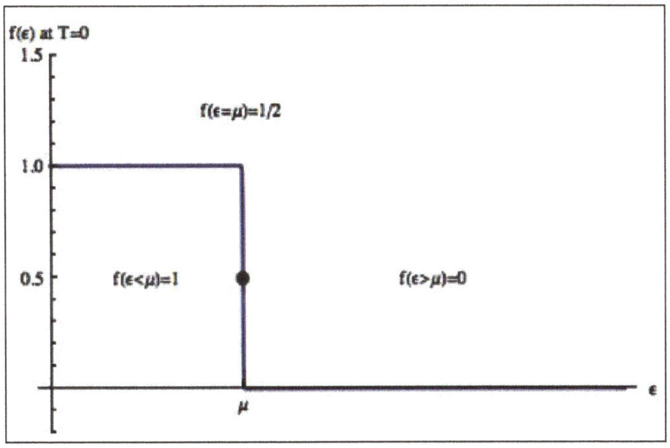

How do we understand this answer for the distribution of occupied states at T = 0? At T = 0 the system is in its groundstate. The groundstate of N free fermions is constructed by filling the N lowest-energy single-particle states. The highest energy level that's filled is called F, the Fermi energy. (Note that this is a single-particle energy level, not energy of all N particles.) The answer we've found for f says that, at T = 0, all states below the chemical potential are filled (f = 1) and all states above the chemical potential are empty (f = 0). So,

$$\epsilon_F = \mu(T=0)$$

Note that μ will be T-dependent.

The value of ϵ_F is determined by N:

$$N = \sum_\alpha \langle n_\alpha \rangle = \int_0^\infty d\epsilon\, D(\epsilon) f(\epsilon) \quad \text{at any } T$$

$$= \int_0^{\epsilon_F = \mu(T=0)} d\epsilon\, D(\epsilon) \quad \text{at } T = 0.$$

$D(\epsilon)$ is the same kind of object that we considered earlier in the context of blackbody radiation.

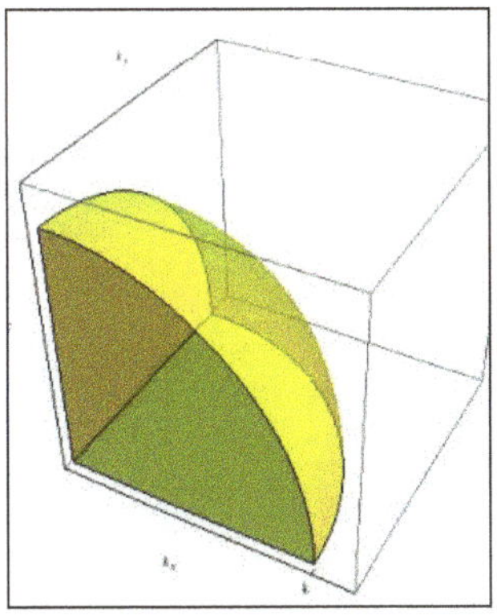

$$N(k) = \#\text{ of states with } |\vec{k}| < k = \underbrace{2}_{\text{spin-}\frac{1}{2}\text{ fermions}} \frac{\frac{1}{8}\frac{4}{3}\pi k^3}{(\pi/L)^3}.$$

For spin-$\frac{1}{2}$ fermions, there are two spin states ('up' and 'down') for each wavenumber.

The only difference is that now we will consider nonrelativistic fermions, which have a dispersion relation:

$$\in = \frac{p^{-2}}{2m} = \frac{\hbar^2}{2m}\vec{k}^2.$$

$$N(\in) = \frac{V}{3\pi^2}\left(\frac{2m}{\hbar^2}\right)^{3/2} \in^{3/2}$$

$$D(\in) = \frac{d}{d\in}N(\in) = V \underbrace{\frac{1}{2\pi^2}\left(\frac{2m}{\hbar^2}\right)^{3/2}}_{\equiv a} \in^{1/2}$$

$$N = N(\in_F) = \int_0^{\in_F} D(\in)d\in = \frac{2}{3}aV \in_F^{3/2} = \frac{2}{3}aV\mu_0^{3/2}$$

This gives the relationship between N and $\mu(T=0)$:

$$\mu(T=0) = \in_F = \left(\frac{N}{V}\frac{3}{2a}\right)^{2/3}$$

The following formula will be useful below:

$$D(\in_F) = \frac{3}{2}\frac{N}{\in_F}.$$

Another useful quantity to define is the Fermi momentum (or Fermi wavenumber):

$$k_F \equiv \sqrt{2m\in_F / \hbar^2}$$

in terms of which:

$$N = 2\frac{\frac{1}{8}\frac{4}{3}\pi k_F^3}{(\pi/L)^3}$$

and so,

$$k_F = \left(3\pi^2\frac{N}{V}\right)^{1/3}.$$

At T = 0, all states with $k < k_F$ (and hence $\in < \in_F$) are filled. All states with $k > k_F$ (and hence $\in > \in_F$) are empty. The locus { $\vec{k}$ such that $|\vec{k}| = k_F$} is called the Fermi surface. The stuff inside is called the Fermi sea.

Now, what's the total energy E of the N fermions in thermal equilibrium?

$$E = \sum_\alpha \in_\alpha \langle n_\alpha \rangle \overset{\text{any } T}{=} \int_0^\infty d\in \underbrace{\in}_{\text{what are we averaging}} \cdot \underbrace{D(\in)}_{\text{how many states have this energy?}} \cdot \underbrace{f(\in)}_{\text{how many are occupied?}}$$

$$\overset{\text{at } T=0}{=} \int_0^{\in_F} d\in \in D(\in) = \int_0^{\in_F} d\in \in V\alpha \in^{1/2}$$

$$\Rightarrow \quad E = \frac{2}{5} aV \in_F^{5/2}.$$

Using the expression $N = N(\in_F) = \int_0^{\in_F} D(\in) d\in = \frac{2}{3} aV \in_F^{3/2} = \frac{2}{3} aV \mu_0^{3/2}$ to eliminate a:

$$E = \frac{3}{5} N \in_F \propto N \left(\frac{N}{V}\right)^{2/3}$$

Notice how completely different this is from a classical ideal gas. In particular, the internal energy depends on the volume. This is because the wavefunctions of the fermions are spread over the container, so they care how big it is.

Electrons in a Metal

How big is $\in_F$ for the conduction electrons in Copper?

$$\frac{N}{V} = \frac{1\,\text{e}^- \text{ per Cu atom}}{1 \text{ Cu atom per unit cell of crystal}} = 10^{22} \cdot 8.45 \frac{e^-}{cm^3}$$

$$\Rightarrow k_F = \left(3\pi^2 \frac{N}{V}\right)^{1/3} = 10^8 \cdot 1.36 cm^{-1}$$

$$\Rightarrow \in_F = \frac{\hbar^2 k_F^2}{2m_e} \approx 7eV.$$

If we define the Fermi velocity v_F by:

$$\in_F = \frac{1}{2} m v_F^2 \quad \Rightarrow v_F = 10^8 \frac{cm}{s} \approx \frac{c}{200}$$

which is much bigger than the few hundred meters per second of classical ideal gas at T_{ROOM} (though they are still reliably non-relativistic). Even though T = 0 here, some of the electrons have lots of kinetic energy. This shows in dramatic fashion that in general you cannot think of temperature as "average kinetic energy" like you can for a classical ideal gas.

We can also define a Fermi temperature by $k_F\, T_F = \in_F$, which comes out to $T_F \approx$ 80, 000K. What's this? It's the temperature that the conduction electrons consider high. Notice that it's bigger than the melting temperature of the solid; actually, it's even bigger than the temperature at which the solid vaporizes. So it's very high. This means that it's a good idea to treat the conduction electrons in a metal as if they are at low temperature for most purposes.

Nonzero (but low) Temperature

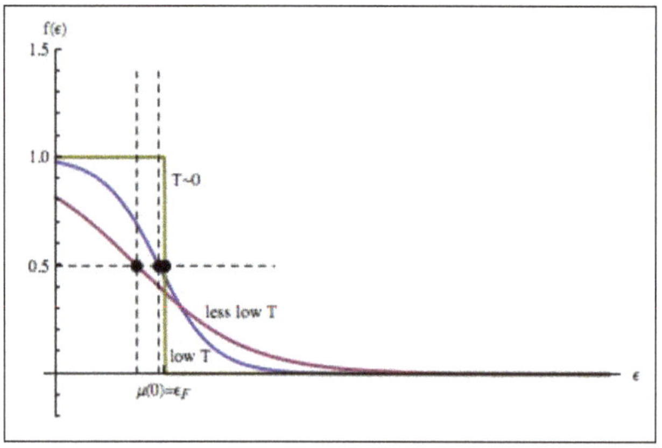

Shown here is the Fermi-Dirac distribution:

$$f(\in) = \frac{1}{e^{(\in-\mu(T))/k_BT}+1}$$

for three values of T, at the same N. Recall that $f(\in = \mu(T)) = \frac{1}{2}$ for any T; the three black dots show the values of $\in = \mu(T)$ where this occurs at each T. μ is a decreasing function of T. Why is that?

At T increases, we fill some states above μ_o and empty some states below μ_o.

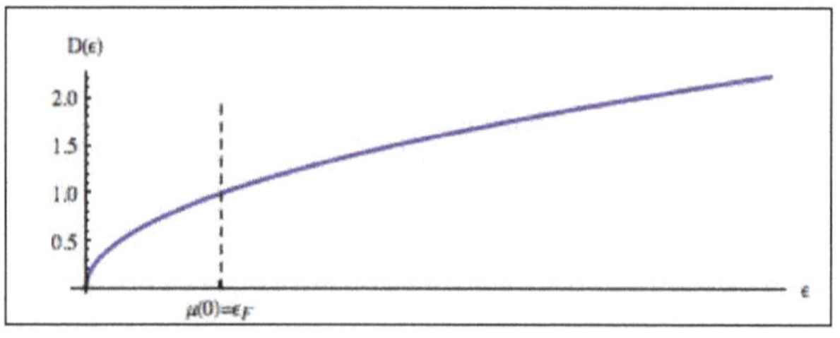

But since D($\in$) is an increasing function of $\in$, μ must decrease in order to keep N fixed.

Classical Ideal Monatomic Gas

The ideal gas is composed of non-interacting atoms. A monatomic gas can be considered to consist of mass points which have linear kinetic energy but no rotational kinetic energy, no vibrational energy and no internal excitations. Thus the total energy of an N particle ideal monatomic gas is:

$$E = \sum_{i=1}^{N} \frac{\vec{p}_i \cdot \vec{p}_i}{2m}$$

The classical partition function is:

$$Z' = \frac{1}{h_o^{3N}} \int \exp\left\{-\beta\left[\frac{1}{2m}\left(\vec{p}_1^{\,2} + \vec{p}_2^{\,2} + \ldots + \vec{p}_N^{\,2}\right)\right]\right\} d\vec{p}_1 \ldots d\vec{p}_N d\vec{q}_1 \ldots d\vec{q}_N$$

The reason for the prime is that this partition function is slightly defective as we shall see. The integrals over the $\vec{q}_1$'s gives a factor of V^N. The integrals over momenta factorize into N separate integrals. So we can write:

$$Z' = \frac{V^N}{h_o^{3N}} \int e^{-\beta \vec{p}_1^2/2m} d^3 p_1 \cdots \int e^{-\beta \vec{p}_N^2/2m} d^3 p_N$$
$$= \zeta^N$$

where,

$$\zeta \equiv \frac{V}{h_o^3} \int_{-\infty}^{\infty} e^{-\beta \vec{p}^2/2m} d^3 p$$

is the partition function for a single molecule. So,

$$\ln Z' = N \ln \zeta$$

It is straightforward to evaluate ζ:

$$\zeta = \frac{V}{h_o^3} \int_{-\infty}^{\infty} e^{-(\beta/2m)\vec{p}^2}\, d^3 p$$
$$= \frac{V}{h_o^3} \int\int_{-\infty}^{\infty} e^{-(\beta/2m)(p_x^2 + p_y^2 + p_z^2)} dp_x dp_y dp_z$$
$$= \frac{V}{h_o^3} \left[\int_{-\infty}^{\infty} e^{-\beta p^2/2m} dp \right]^3$$

Recall Reif:

$$\int_{-\infty}^{\infty} e^{-\alpha x^2} dx = \sqrt{\frac{\pi}{\alpha}}$$

So we evaluate the integral:

$$\int_{-\infty}^{\infty} e^{-\beta p^2/2m} dp = \sqrt{\frac{\pi 2m}{\beta}}.$$

Hence,

$$\zeta = V \left(\frac{2\pi m}{h_o^2 \beta} \right)^{3/2}$$

So,

$$Z' = \zeta^N = \left[V \left(\frac{2\pi m}{h_o^2 \beta} \right)^{3/2} \right]^N$$

and,

$$\ln Z' = N\left[\ln V - \frac{3}{2}\ln\beta + \frac{3}{2}\ln\left(\frac{2\pi m}{h_o^2}\right)\right].$$

From our previous equations we can calculate the mean pressure $\bar{p}$, the mean energy $\bar{E}$, and the specific heat C_V:

$$\bar{p} = \frac{1}{\beta}\frac{\partial \ln Z'}{\partial V} = \frac{1}{\beta}\frac{N}{V}$$

or,

$$\bar{p}V = Nk_BT$$

This is the equation of state of the ideal gas. The mean energy is given by:

$$\bar{E} = -\frac{\partial}{\partial\beta}\ln Z' = \frac{3}{2}\frac{N}{\beta}$$

or,

$$\bar{E} = \frac{3}{2}Nk_BT$$

This is what we expect from the equipartition theorem. The heat capacity is given by:

$$C_V = \left(\frac{\partial \bar{E}}{\partial T}\right)_V = \frac{3}{2}Nk_B$$

Using $N = \nu N_a$, where N_a is Avogadro's number and ν is the number of moles, and $R = N_a k_B$, we get for the molar specific heat:

$$c_V = \frac{3}{2}R$$

This agrees with our previous result.

We can also calculate the entropy S:

$$S = k_B\left[\ln Z' + \beta\bar{E}\right] = Nk_B\left[\ln V - \frac{3}{2}\ln\beta + \frac{3}{2}\ln\left(\frac{2\pi m}{h_o^2}\right) + \frac{3}{2}\right]$$

or,

$$S = Nk_B\left[\ln V + \frac{3}{2}\ln T + \sigma\right]$$

where,

$$\sigma = \frac{3}{2}\ln\left(\frac{2\pi m k_B}{h_o^2}\right) + \frac{3}{2}$$

is a constant independent of *T*, *V*, and *N*. This expression for the entropy is wrong.

Gibbs Paradox

To see why $\sigma = \frac{3}{2}\ln\left(\frac{2\pi m k_B}{h_o^2}\right) + \frac{3}{2}$ is wrong, note that as $T \to 0,\ S \to -\infty$ instead of $S \to 0$. This violates the third law of thermodynamics. But this is not surprising since classical statistical mechanics breaks down as $T \to 0$. At low temperatures one must use quantum mechanics. Recall that $S = k_B \ln \Omega$ where Ω is the number of accessible microstates. As $T \to 0$, the only accessible microstate is the ground state. In addition the spacing of the microstates becomes larger than the temperature $(\Delta E > k_B T)$, and we must use quantum mechanics to address the discrete energy levels. So it's not surprising that S doesn't work at low temperatures.

But there's another problem with *S*. Namely the entropy *S* does not scale with the size of the system as an extensive parameter should. In other words if V and N double, then the S and E should double. Consider a box with an internal partition. The total entropy is $S_1 + S_2$. We know this to be true since $Z = Z_1 Z_2$ if systems 1 and 2 are completely independent, as they are in this example. Thus,

$$S_1 = k_B\left[\ln Z_1 + \beta \bar{E}_1\right]$$
$$S_2 = k_B\left[\ln Z_2 + \beta \bar{E}_2\right]$$

and,

$$\begin{aligned} S &= k_B\left[\ln Z_1 Z_2 + \beta\left(\bar{E}_1 + \bar{E}_2\right)\right] \\ &= k_B\left[\ln Z_1 + \beta \bar{E}_1\right] + k_B\left[\ln Z_2 + \beta \bar{E}_2\right] \\ &= S_1 + S_2 \end{aligned}$$

Now remove the partition. If the gases on the two sides are identical ($N_1 = N_2, V_1 = V_2$, and $\bar{E}_1 = \bar{E}_2$ and $S_1 = S_2$), then we will observe no physical changes. Nothing happens. Thus we would expect:

$$S = S_1 + S_2 = 2S_1$$

If the two sides are identical.

Now go back to our expression for the entropy of an ideal gas and note that:

$$\begin{aligned} S(2N, 2V) - 2S(N, V) &= 2Nk_B\left[\ln(2V) + \frac{3}{2}\ln T + \sigma\right] - 2Nk_B\left[\ln V + \frac{3}{2}\ln T + \sigma\right] \\ &= 2Nk_B \ln 2 \\ &\neq 0 \end{aligned}$$

The fact that $S(2N, 2V) - 2S(N, V) \neq 0$ means something is wrong because the entropy is not additive. This problem is known as the "Gibbs paradox."

If the gases on the two sides were different, then an increase in S when the partition function is removed seems reasonable since the two gases will diffuse and intermingle. In other words we expect the configurational entropy to increase. Then putting back the partition clearly would not restore the total system to its original state. On the other hand, if the gases are identical, then the restoration of the partition returns the system to a state physically indistinguishable from the initial state.

So if the gases on the two sides are identical, this increase in S does not make physical sense. The difficulty is associated with the concept of indistinguishability of the particles. If we exchange the positions of two identical particles, then we really haven't changed any physical properties of the system. Yet we have counted these two identical microstates as truly distinct in calculating the entropy. (Recall $S = k_B \ln \Omega$).

This concept of indistinguishability is clearly well established in quantum mechanics where this concept is basic to the theory. We have taken the classical picture of atoms too seriously and have finally exceeded the limits of classical validity.

Gibbs postulated that the classical partition function should be divided by N! to correct for counting each indistinguishable microstate. Let,

$$Z = \frac{Z'}{N!}$$

Then,

$$\ln Z = \ln Z' - N \ln N + N$$

Where Stirling's approximation has been used. The entropy becomes:

$$S = k_B\left[\ln \mathrm{Z} + \beta \bar{\mathrm{E}}\right]$$

$$= k_B N\left[\ln \mathrm{V} + \frac{3}{2}\ln \mathrm{T} + \sigma\right] + k_B\left[N - N \ln N\right]$$

$$= k_B N\left[\ln \frac{V}{N} + \frac{3}{2}\ln T + \sigma + 1\right]$$

or

$$S = k_B N\left[\ln \frac{\mathrm{V}}{\mathrm{N}} + \frac{3}{2}\ln \mathrm{T} + \sigma_o\right]$$

Where,

$$\sigma_o = \sigma + 1 = \frac{3}{2}\ln\left(\frac{2\pi m k_B}{h_o^2}\right) + \frac{5}{2}$$

Now when the system is doubled ($N \to 2N, V \to 2V$), the entropy is doubled. Later when we repeat the calculation of the entropy of an ideal monatomic gas using quantum statistics, the factor of N! will arise naturally. Also the constant h_o will be identified as Planck's constant.

It is interesting to note that this basic concept of quantum mechanics (indistinguishability of identical particles) did not arise spontaneously with the development of quantum mechanics, but was suggested by the ad hoc resolution of a paradox in classical statistical mechanics. Thus the Gibbs' paradox joins black body radiation, the photoelectric effect, atomic line spectra and all the other pieces of evidence that led theorists to advance beyond classical mechanics to quantum mechanics.

Validity of the Classical Approximation

We have used a classical approximation to describe an ideal monatomic gas. Under what conditions would we expect it to be valid? One approach to answer this question begins with the uncertainty relation.

$$\Delta x \Delta p \gtrsim \hbar$$

We expect $\Delta x \lesssim \bar{R}$ where $\bar{R}$ is the mean separation of the ideal gas atoms, and $\Delta p \lesssim \bar{p}$ where $\bar{p}$ is the mean momentum. Then we would expect that classical mechanics would adequately describe the dynamics of the gas whenever,

$$\bar{R}\bar{p} \gg \hbar$$

This effectively says that $\hbar$ is negligible. Recall that:

$$\bar{p} = \frac{2\pi\hbar}{\bar{\lambda}} = \frac{\hbar}{\bar{\lambda}}$$

where $\bar{\lambda}$ is the mean de Broglie wavelength. Then:

$$\bar{R}\bar{p} \gg \hbar$$

Becomes:

$$\bar{R} \gg \frac{\hbar}{\bar{p}} = \frac{\bar{\lambda}}{2\pi}$$

Or more simply, the classical description applies when:

$$\bar{R} \gg \bar{\lambda}$$

i.e., the mean separation between particles is much greater than their mean de Broglie wavelength.

To estimate $\bar{R}$, imagine each molecule at the center of a little cube of side $\bar{R}$, these cubes filling the available volume V. Then:

$$\bar{R}^3 N = V$$

or

$$\overline{R} = \left(\frac{V}{N}\right)^{1/3}$$

The mean momentum $\overline{p}$ can be estimated using the equipartition theorem:

$$\frac{\overline{p}^2}{2m} \approx \frac{3}{2} k_B T$$

or

$$\overline{p} \approx \sqrt{3mk_B T}$$

and,

$$\overline{\lambda} \approx \frac{h}{\sqrt{3mk_B T}}$$

Hence the condition that the classical description should be valid is:

$$\left(\frac{V}{N}\right)^{1/3} \gg \frac{h}{\sqrt{3mk_B T}}$$

So for large masses and high temperatures we expect classical mechanics to hold.

As an example, air at standard temperature and pressure (STP) has T = 300 K and mean pressure $\overline{P}$ = 760 mm of Hg (mercury)$= 1.01 \times 10^5$ Newton/ m^2. 80% of air is N_2. So the mass of a nitrogen molecule is $m = 28m_p$ where $m_p = 1.67 \times 10^{-27}$ kilograms is the mass of a proton. Using $k_B = 1.38 \times 10^{-23}$ Joules/K, we obtain:

$$\frac{N}{V} = \frac{\overline{p}}{k_B T} = 2.4 \times 10^{25} \text{ molecules/m}^3$$

or,

$$\overline{R} = \left(\frac{V}{N}\right)^{1/3} = 35\mathring{A}$$

and,

$$\overline{\lambda} = \frac{h}{\overline{p}} = \frac{h}{\sqrt{3mk_B T}} \cong 0.3\mathring{A}$$

Thus,

$$\overline{R} \gg \overline{\lambda}$$

So the classical description holds for air. However, this is not the case for an electron gas in a metal at 300 K. In that case we find $\bar{R} \cong 2$ or 3 $\mathring{A}$ and $\bar{\lambda} \cong 60\ \mathring{A}$ (using the electron mass). Thus $\lambda \gg R$ and we must use a full quantum treatment for an electron gas.

Classical Theory of the Specific Heat of Solids

Consider a monatomic solid as a collection of N independent, identical atoms of mass m, all of which vibrate harmonically about their equilibrium positions at frequency ω. The total energy of the solid is:

$$E = \sum_{i=1}^{3N} \left(\frac{p_i^2}{2m} + \frac{1}{2} k_i q_i^2 \right)$$

$$= \sum_{i=1}^{3N} \left(\frac{p_i^2}{2m} + \frac{1}{2} k q_i^2 \right)$$

If all the spring constants k are equal. By the equipartition theorem are equal. By the equipartition theorem:

$$\bar{E} = (\text{number of quadratic terms in } E) \times \frac{1}{2} k_B T$$

$$= 3Nk_B T$$

and the molar specific heat becomes simply:

$$c_V = \left. \frac{\partial \bar{E}}{\partial T} \right|_V = 3N_A k_B = 3R$$

where N_A is Avogadro's number ($N_A = 6.02 \times 10^{23}$ molecules/mole) or:

$$c_V = 25 \text{ Joules/ mole } - \text{K}$$

Notice that the specific heat is a constant independent of T. This is is called the Dulong– Petit value of the specific heat. The values of c_p for some common substances at room temperature in units of Joules mole^{-1} K^{-1} are:

Solid	C_p	Solid	c_p
Cu	24.5	Al	24.4
Ag	25.5	Sn	26.4
Pb	26.4	S	22.4
Zn	25.4	C (diamond)	6.1

Note the obvious discrepancy with carbon (diamond). The classical theory predicts a temperature independent specific heat while all materials have the general property that c_V is relatively temperature independent at high temperature but goes to zero as $T \to 0$ as required by the third law of thermodynamics. From the graph one can see that above about 1000 K, diamond has a specific heat of approximately 25 Joules/mole–K.

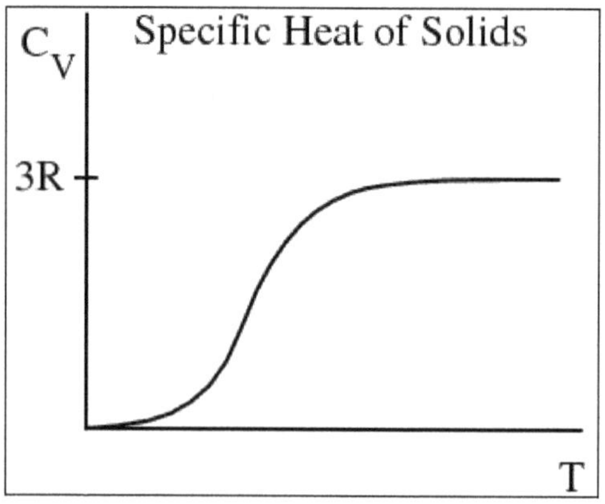

Einstein Model of Specific Heat

To get the proper temperature dependence of the specific heat of solids, we must use quantum mechanics to take into account the discrete energy levels that become apparent at low energies and low temperatures. One of the simplest ways to do this is provided by the Einstein model of specific heat. As a simple model Einstein proposed a collection of N independent, identical harmonic oscillators but treated the oscillators in quantum mechanical rather than classical terms. Let the spring constant $k_i = m\omega_i^2 = m\omega^2 =$ constant for all atoms in the monatomic solid. Then:

$$E = \sum_{i=1}^{3N} \left(\frac{p_i^2}{2m} + \frac{1}{2} m\omega q_i^2 \right)$$

The partition function for N identical independent harmonic oscillators is:

$$Z = \zeta^{3N}$$

where,

$$\zeta = \sum_n e^{-\beta \epsilon_n}$$

The energy levels of a single one dimensional harmonic oscillator are:

$$\epsilon_n = \hbar\omega \left(n + \frac{1}{2} \right)$$

where n = 0, 1, 2, ... Then:

$$\zeta = \sum_{n=0}^{\infty} e^{-\beta\hbar\omega(n+1/2)} = e^{-\beta\hbar\omega/2} \sum_{n=0}^{\infty} e^{-\beta\hbar\omega n} = e^{-\beta\hbar\omega/2} \left(1 + e^{-\beta\hbar\omega} + e^{-2\beta\hbar\omega} + \ldots \right)$$

The sum is just an infinite geometric series where each term is obtained from the preceding one as a result of multiplication by $e^{-\beta\hbar w}$. We can easily sum the series to get:

$$\zeta = e^{-\beta\hbar\omega/2}\,\frac{1}{1-e^{-\beta\hbar\omega}}$$

or,

$$\ln\zeta = -\frac{1}{2}\,\beta\hbar\omega - \ln\left(1-e^{-\beta\hbar w}\right)$$

and,

$$\ln Z = 3N\ln\zeta = -\frac{3}{2}N\beta\hbar\omega - 3N\;\ln\left(1-e^{-\beta\hbar\omega}\right)$$

We can now obtain the mean energy:

$$\begin{aligned}\bar{E} &= -\frac{\partial\ln Z}{\partial\beta}\\ &= -3N\frac{\partial}{\partial N}\ln\zeta\\ &= \frac{3N\hbar\omega}{2} + \frac{3N\hbar\omega e^{-\beta\hbar\omega}}{1-e^{-\beta\hbar\omega}}\\ &= 3N\hbar\omega\left(\frac{1}{2}+\frac{1}{e^{-\beta\hbar\omega}-1}\right)\end{aligned}$$

And the heat capacity:

$$\begin{aligned}C_V &= \left.\frac{\partial\bar{E}}{\partial T}\right|_V\\ &= \left.\frac{\partial\bar{E}}{\partial\beta}\right|_V\frac{\partial\beta}{\partial T}\\ &= -\frac{1}{k_BT^2}\left(\frac{\partial\bar{E}}{\partial\beta}\right)_V\\ &= -\frac{3N\hbar\omega}{k_BT^2}\left[-\frac{e^{-\beta\hbar\omega}\hbar\omega}{\left(e^{\beta\hbar\omega}-1\right)^2}\right]\\ &= 3Nk_B\left(\frac{\hbar\omega}{k_BT}\right)^2\frac{e^{\beta\hbar\omega}\hbar\omega}{\left(e^{\beta\hbar\omega}-1\right)^2}\end{aligned}$$

It is customary to define a quantity called the Einstein temperature θ_E by:

$$\beta\hbar\omega = \frac{\hbar\omega}{k_B T} \equiv \frac{\theta_E}{T}$$

or,

$$\theta_E \equiv \frac{\hbar\omega}{k_B}$$

Then the molar specific heat becomes (with $N = N_A$):

$$c_V = 3N_A k_B \left(\frac{\theta_E}{T}\right)^2 \frac{e^{\theta_E/T}}{\left(e^{\theta_E/T}-1\right)^2}$$

$$= 3R\left(\frac{\theta_E}{T}\right)^2 \frac{e^{\theta_E/T}}{\left(e^{\theta_E/T}-1\right)^2}$$

In the high temperature limit of $T \to \infty$ or $\theta_E / T \to 0$, we can expand the exponential in the denominator:

$$\left(e^{\theta_E/T}-1\right)^2 \to \left(1+\frac{\theta_E}{T}-1\right)^2 = \left(\frac{\theta_E}{T}\right)^2$$

and,

$$c_V \to 3R \text{ as } T \to \infty$$

In the low temperature limit where $T \to 0$ or $\theta_E / T \to \infty$

$$c_V \to 3R\left(\frac{\theta_E}{T}\right)^2 e^{-\theta_E/T} \text{ as } T \to 0$$

Careful experiments indicate that c_V goes as T^3 at low temperatures rather than exponentially. The explanation will come later when we consider the Debye theory of collective oscillations or normal modes of a crystalline solid consisting of a lattice of atoms.

Ideal Bose Gas

Boson-In quantum mechanics, a boson is a particle that follows Bose–Einstein statistics.

Equation of State

We consider a gas of non-interacting bosons in a volume V at temperature T and chemical potential μ. The system is allowed to interchange particles and energy with the surroundings. The appropriate ensemble to treat this many-body system is the grand canonical ensemble.

Non-relativistic Bosons: Our bosons are non-relativistic particles with spin s, whose one-particle energies $\in$ (k),

$$\in (\mathrm{k}) = \in (k) = \frac{\hbar^2 k^2}{2m}, \qquad \in_0 = \in (0) = 0$$

include only the kinetic energy term.

Negative Chemical Potential

The chemical potential obeys,

$$-\infty < \mu < \in_0, \qquad \in_0 = 0.$$

A chemical potential larger than the lowest energy state would lead to nonphysical level occupation.

$$n(\in_r) = \langle \hat{n}_r \rangle \ \frac{1}{e^{\beta(\in_r - \mu)} - 1}.$$

Approaching the Thermodynamic Limit

We consider a situation when the gas is in a box with volume $V = L_x L_y L_z$ and subject to periodic boundary conditions.

In the thermodynamic limit ($N \to \infty$, $V \to \infty$, with $n = N/V = \text{const}$), the sums over the wave vector $\vec{k}$ can be replaced by integrals as in the case of the Fermi gas.

However, here we have to be careful when μ happens to approach the value 0. In order to see what kind of trouble we then might get into, let us calculate the ground state occupation.

Occupation of the Lowest Energy State

We consider the expectation value of the ground state for μ approaching zero from below, viz when $-\beta\mu \ll 1$:

$$\mathrm{n}(\in_0) = \frac{1}{e^{-\beta\mu} - 1} = \frac{1}{(1 - \beta\mu + \ldots) - 1} \approx -\frac{1}{\beta\mu}, \qquad \in_0 = 0,$$

which means that $n(\in_0) = \langle \hat{n}_{r=0} \rangle$ diverges. The lowest energy state may hence by occupied macroscopically. This is the case when:

$$\frac{1}{|\mu|\beta} \sim N, \qquad |\mu| \sim \frac{k_B T}{N}, \qquad 1 - z \sim \frac{1}{N}.$$

Density of States

The density of states D(E), which has the same expression as for fermionic systems:

$$D(E) \sim \sqrt{E}, \qquad \lim_{E \to 0} D(E) = 0$$

vanishes for $E \to 0$. This is where we are going to encounter a problem, if we replace:

$$\frac{1}{V}\sum_{r} \to \int dE\, D(E)$$

we will get that the ground state has zero weight even though, as we have just shown that it can be macroscopically occupied. Fermionic systems do not encounter this problem due to the Pauli principle, which imposes that $\langle \hat{n}_r \rangle \in [0,1]$.

Special Treatment for the Ground State

The problem with the potentially macroscopic occupation of the ground state can be solved by giving it.

$$\begin{aligned}\beta\Omega(T,\, V,\, z) &= \sum_{r} \ln\left[1 - e^{-\beta(\epsilon_r - \mu)}\right] \\ &= (2s\,+\,1)\frac{V}{(2\pi)^3}4\pi \int_0^{\infty} dk\, k^2 \ln\,\left(1 - ze^{-\beta\epsilon(k)}\right) \\ &+\ \underbrace{(2s\,+\,1)\ \ln(1\,-\,z)}_{\text{occupation of the ground state } \epsilon(0)=0}\ .\end{aligned}$$

A special treatment: Starting from the expression for the bosonic grand canonical potential $\Omega(T,\, V,\, \mu)$ we have split the $\sum_r$ into an integral over all states, for the case of fermions, and into the ϵ_0 contribution (the last term).

Equation of State

Irrelevance of Condensate

The ground state contribution to the grand canonical potential Ω is formally irrelevant in the thermodynamic limit as a consquence of the scaling of the chemical potential:

$$\lim_{V\to 0} \frac{\ln(1-z)}{V} \approx \lim_{V\to 0} \frac{-\ln(N)}{V} \to 0.$$

We note, however, that the size of the condensate, that is the number of particles occupying the ground state, determines how many particles occupy energiers $E > E_0$, viz the density of the normal fluid.

Dimensionless Variables

With the dimensionless variable x and the thermal de Broglie wavelength λ,

$$x = \hbar k\sqrt{\frac{\beta}{m}}, \qquad \lambda = \sqrt{\frac{2\pi\beta\hbar^2}{m}},$$

we write as:

$$\beta\Omega(T,\ V,\ z) = \frac{2s+1}{\lambda^3}\frac{4V}{\pi}\int_0^\infty dx\, x^2\ \ln\left(1 - ze^{-x^2}\right),$$

All in parallel to the transformations performed for the Fermi gas. Compare equation:

$$-\beta\,\Omega(T,\ V,\ \mu) = \frac{(2s+1)}{\lambda^3}\ \frac{4V}{\sqrt{\pi}}\int_0^\infty dx\, x^2 \ln\left(1 + z\, e^{-x^2}\right).$$

Taylor Expansion

We recall that the Taylor series expansion:

$$\ln(1-y) = -\sum_{n=1}^{\infty}\frac{y^n}{n}, \qquad |y| < 1$$

may be used, to express the integral:

$$\int_0^\infty dx\, x^2\ \ln\left(1 - ze^{-x^2}\right) = -\frac{\sqrt{\pi}}{4}\sum_{n=1}^{\infty}\frac{z^n}{n^{5/2}}$$

in terms of,

$$g_{5/2} = -\frac{4}{\sqrt{\pi}}\int_0^\infty dx\, x^2\ \text{In}\left(1 - ze^{-x^2}\right) = \sum_{n=1}^{\infty}\frac{z^n}{n^{5/2}}.$$

Note that $g_{5/2}(z)$ and $f_{5/2}(z)$, as defined in, differ by a sign $(-1)^{n+1}$ in the summand. For later uses we also define $g_{3/2}(z)$ as:

$$g_{3/2}(z) = z\frac{d}{dz}g_{5/2}(z) = \sum_{n=1}^{\infty}\frac{z^n}{n^{3/2}}.$$

Note that $f_{3/2}(z)$ was defined analogously in as $zd\left(f_{5/2}\right)/dz$.

Bosonic Grand Canonical Potential

With, the grand canonical potential takes the form:

$$\beta\Omega(T,\ V,\ z) = -\frac{2s+1}{\lambda^3}\ Vg_{5/2}(z)$$

Except for the extra term on the right-hand side, and for an exchange $g_{5/2} \leftrightarrow f_{5/2}$, it has the same form as the expression for the Fermi gas.

1. Pressure:

From $\Omega = -PV$ and we get:

$$\beta P = \frac{2s+1}{\lambda^3} g_{5/2}(z)$$

2. Particle density:

For the particle density N/V the relation:

$$\frac{\langle N \rangle}{V} = \frac{1}{\beta V}\left[\frac{\partial}{\partial z} \ln \mathcal{Z}\right] = \frac{z}{V}\left[\frac{\partial}{\partial z} \ln \mathcal{Z}\right] = -\frac{\beta z}{V}\left[\frac{\partial}{\partial z}\Omega\right]$$

$\beta\Omega = -\ln \mathcal{Z}$, in this case the condensate term contributes. We then find:

$$n = \frac{2s+1}{\lambda^3} g_{3/2}(z) + \frac{2s+1}{V}\frac{z}{1-z}$$

For the density of particles $n = \langle \hat{N} \rangle / V$ in terms of the fugacity z, where we have used $\beta\Omega(T, V, z) = -\frac{2s+1}{\lambda^3} V g_{5/2}(z)$ and $g_{3/2}(z) = z\frac{d}{dz} g_{5/2}(z) = \sum_{n=1}^{\infty} \frac{z^n}{n^{3/2}}$. The thermal equation of states then results from combining $\beta P = \frac{2s+1}{\lambda^3} g_{5/2}(z)$ and $n = \frac{2s+1}{\lambda^3} g_{3/2}(z) + \frac{2s+1}{V}\frac{z}{1-z}$.

Ground State Occupation

The term:

$$\frac{2s \quad 1}{V} \frac{z}{z}$$

In $n = \frac{2s+1}{\lambda^3} g_{3/2}(z) + \frac{2s+1}{V}\frac{z}{1-z}$ describes The contribution of the ground state to the particle density n. When n_0 becomes macroscopically large on speaks of a Bose-Einstein condensation.

Internal Energy

The internal energy U is given by:

$$U = -\left(\frac{\partial}{\partial \beta} \ln \mathcal{Z}\,(T, z, V)\right)_{z,V} = \left(\frac{\partial}{\partial \beta} \beta\Omega(T, V, z)\right)_{z,V}$$

Note that the fugactity z is kept constant in above expression. Our result states that:

$$\beta\Omega(T, V, z) \sim -\lambda^{-3} \sim -\beta^{-3/2},$$

which then leads with equation above to:

$$\frac{U}{V}=\frac{3k_BT}{2}\frac{2s+1}{\lambda^3}g_{5/2}(z)$$

This expression is the same as the one for the Fermi gas when $f_{5/2}(z)$ is substituted by $g_{5/2}(z)$. The reason is that the ground state energy ϵ_0 vanishes, $\epsilon_0=0$. It does hence not matter how many particle occupy the lowest energy level.

Caloric Equation of State

Combining $\frac{U}{V}=\frac{3k_BT}{2}\frac{2s+1}{\lambda^3}g_{5/2}(z)$ with $\beta P=\frac{2s+1}{\lambda^3}g_{5/2}(z)$ one can derive the caloric equation of state:

$$U=\frac{3}{2}PV,$$

which is identical to the one obtained for the ideal Fermi gas.

Classical Limit

The classical limit (non-degenerate Bose gas) corresponds to low particle densities and high temperatures. The fugacity is then small,

$$z=e^{\beta\mu}\ll 1,$$

with the Bose-Einstein distribution:

$$\langle\hat{n}_r\rangle=\frac{1}{z^{-1}e^{\beta\epsilon_r}-1}=\frac{ze^{-\beta\epsilon_r}}{1-ze^{-\beta\epsilon_r}}\approx ze^{-\beta\epsilon_r}\ll 1$$

Reducing to the Maxwell-Boltzmann distribution, just as for a fermionic system. The differences between Bose-, Fermi- and Boltzmann statistics are in next order of the order 1/z and hence small.

Expansion in the Fugacity

As $z\ll 1$, it is sufficient to retain only the first two terms of the series for $g_{5/2}(z)$ and $g_{3/2}(z)$:

$$g_{5/2}(z)\approx z+\frac{z^2}{2^{5/2}},\qquad g_{3/2}(z)\approx z+\frac{z^2}{2^{3/2}}.$$

With that, the particle density $n=\frac{2s+1}{\lambda^3}g_{3/2}(z)+\frac{2s+1}{V}\frac{z}{1-z}$ takes the following form:

$$n\approx\frac{2s+1}{\lambda^3}z\left(1+\frac{z}{2^{3/2}}\right)+\frac{2s+1}{V}\frac{z}{1-z}.$$

Irrelevance of the Ground-state Contribution

We note that the ground-state contribution vanishes generically in the thermodynamic limit, being proportional to 1/V. The number of particles occupying the ground state is finite only for $z \to 1$. For the case of small fugacities considered here we can neglect it generically, obtaining:

$$n \approx \frac{2s+1}{\lambda^3} z \left(1 + \frac{z}{2^{3/2}}\right).$$

Convergence Radius

That ground state contributions can be generically neglected at elevated temperatures follows also from the following consideration.

- An expansion in a physical parameter, like z or β, converges only as long as one one remains within the same phase, here the high-temperature quasi-classical gas phase.
- An expansion diverges once a phase boundary is encountered, in our case the transition to a phase with macroscopically occupied ground-state level. A condensed low-temperature state can therefore not be described within a high-temperature expansion.

Classical Limit

In the strict classical limit we retain only the the terms ~ z on the right-hand side of we thus have that:

$$n\lambda^3 \approx (2s + 1) z^{(0)}, \quad z^{(0)} \approx \frac{n\lambda^3}{2s+1}.$$

in the zeroth approximation.

Classical Equation of State

We note that $n\lambda^3 \approx (2s + 1) z^{(0)}, \quad z^{(0)} \approx \frac{n\lambda^3}{2s+1}$ is identical to corresponding expression for Fermions.

The expression $\beta P = \frac{2s+1}{\lambda^3} g_{5/2}(z)$ for the pressure reduces then with $g_{5/2} \to z^{(0)}$ to the equation of state for classical particles:

$$\beta P = \frac{2s+1}{\lambda^3} \frac{n\lambda^3}{2s+1}, \qquad VP = \langle \hat{N} \rangle k_B T, \qquad n = \langle \hat{N} \rangle / V.$$

1. First order correction:

We solve $n\lambda^3 \approx (2s + 1) z^{(0)}, \quad z^{(0)} \approx \frac{n\lambda^3}{2s+1},$

$$\frac{n\lambda^3}{2s+1} \approx z^{(0)} \approx z^{(1)}\left(1+\frac{z^{(1)}}{2^{3/2}}\right) \approx z^{(1)}\left(1+\frac{z^{(0)}}{2^{3/2}}\right)$$

for $z^{(1)}$, obtaining:

$$z^{(1)} \approx \frac{z^{(0)}}{1+\frac{z^{(0)}}{2^{3/2}}} \approx z^{(0)}\left(1-\frac{z^{(0)}}{2^{3/2}}\right),$$

which may be substituted in the pressure equation for the ideal Bose gas:

$$\beta P \approx \frac{2s+1}{\lambda^3} g_{5/2}(z) \approx \frac{2s+1}{\lambda^3} z(1).$$

2. Quantum correction:

Taking above equations together, gives us:

$$PV = \langle \hat{N} \rangle k_B T \left[1-\frac{n\lambda^3}{4\sqrt{2}(2s+1)}\right].$$

The last term in this expression are the quantum corrections.

1. Equation of states for a real gas, like the van der Waals equation , posses “similar” additive corrections with respect to the ideal case, which are however due to the interaction between particles. The additive terms present in originate on the other side from the indistinguishability principle and not from the interaction among particles.
2. The correcting term for the ideal Fermi gas quasi-classical equation of state is positive, contributing as “repulsion” among particles. For the Bose gas, the additive term is negative and therefore contributes as an ”attraction” among particles.

Quantitatively, the quantum corrections are much smaller than terms coming from the interaction among particles.

Bose-Einstein Condensation

We consider now the limit of high particle densities and low temperatures (quantum limit), where one finds important qualitative differences between bosons, fermions and classical particles.

1. Particle density:

We rewrite the the particle density as:

$$n = \int_0^\infty \frac{D(\in)d\in}{e^{\beta(\in-\mu)}-1} + \frac{2s+1}{V}\frac{z}{1-z},$$

where the the density of states D(∈) is given by:

$$D(\in) = A\sqrt{\in}, \qquad A = \frac{2s+1}{(2\pi)^2}\left(\frac{2m}{\hbar^2}\right)^{3/2}.$$

2. Dimensionless variables:

The regular contribution to the particle density in can be evaluated for μ = 0 as:

$$\lim_{\mu\to 0}\int_0^\infty \frac{A\sqrt{\in}d\in}{e^{\beta(\in-\mu)}-1} = \frac{A}{\beta^{3/2}}\int_0^\infty \frac{\sqrt{x}dx}{e^x-1} = \frac{A}{\beta^{3/2}}\cdot 2.61 ,$$

where we have used the dimensionless variable $x = \beta\in$. The original expression for n then becomes:

$$n = 2.61A(k_BT)^{3/2} + \frac{2s+1}{V}\frac{1}{e^{-\beta\mu}-1}, \quad A = \frac{2s+1}{(2\pi)^2}\left(\frac{2m}{\hbar^2}\right)^{3/2}.$$

This is a mixed representation where we have taken the limes μ → 0 for the regular contribution, but not for the occupation of the ground state.

3. Bose-Einstein condensation:

It is evident from $n = 2.61A(k_BT)^{3/2} + \frac{2s+1}{V}\frac{1}{e^{-\beta\mu}-1}, \quad A = \frac{2s+1}{(2\pi)^2}\left(\frac{2m}{\hbar^2}\right)^{3/2}$ that there is a critical temperature T_c:

$$n = 2.61A(k_BT_c)^{3/2}, \quad n = 2.61\frac{2}{\sqrt{\pi}}\frac{2s+1}{\lambda_c^3}, \quad \lambda_c = \sqrt{\frac{h^2}{2\pi mk_BT_c}}$$

for which the regular contribution would fall below the desired particle density n. We have used that $(4\pi)^{3/2}/4\pi^2 = 2/\sqrt{\pi}$.

- T_c is the Bose-Einstein transition temperature.
- A non-vanishing negative chemical potential μ < 0 would lead to an even small regular term in $n = \int_0^\infty \frac{D(\in)d\in}{e^{\beta(\in-\mu)}-1} + \frac{2s+1}{V}\frac{z}{1-z}$. There is therefore no way that the regular term could account for all particle for $T < T_c$.
- The transition takes place when $n\lambda_c^3/(2s+1) = 2.61\cdot 2/\sqrt{\pi} \approx 2.9$, viz when the thermal wavelength λc is of the order of the inter-particle distance.

a. Scaling of the chemical potential:

Rewriting $n = 2.61A(k_BT)^{3/2} + \frac{2s+1}{V}\frac{1}{e^{-\beta\mu}-1}$, $A = \frac{2s+1}{(2\pi)^2}\left(\frac{2m}{\hbar^2}\right)^{3/2}$ for small $|\mu|$ as:

$$n - n_c(T) \sim \frac{2s+1}{V}\frac{k_BT}{-\mu}, \quad (-\mu) \sim \frac{2s+1}{V}\frac{k_BT}{n-n_c(T)},$$

where $n_c(T) = 2.61(2s+1)/\lambda^3$. The chemical potential scales therefore like 1/V, viz it strictly vanishes only in the thermodynamic limit $V \to \infty$.

b. First excited state:

The energy level are quantized for a particle in a box,

$$\in(\mathrm{k}) = \frac{k_x^2 + k_y^2 + k_z^2}{2m}, \qquad k_\alpha = \frac{2\pi}{L_\alpha}n_\alpha, \qquad \alpha = x, y, z,$$

The volume is with $V = L_xL_yL_z$ the product of the linear dimensions.

c. Diverging occupation of the first excited state:

The energy $\in_1$ of one of the first excited states, corresponding e.g. to $(n_x, n_y, n_z) = (1, 0, 0)$, then scales as:

$$\in_1 \sim \frac{1}{L_x^2} \sim V^{-2/3}, \qquad \in_1 \gg |\mu| \sim V^{-1}.$$

The occupation n_1 of the first excited state:

$$n_1 = \frac{1}{e^{\beta(\in_1-\mu)}-1} \approx \frac{1}{e^{\beta\in_1}-1} \approx \frac{1}{1+\beta\in_1-1} \sim V^{2/3},$$

Therefore diverges in the thermodynamic limit $V \to \infty$. The ground-state occupation $n_0 \sim V$ diverges in contract to a macroscopic value.

Ideal Fermi Gas

The properties of an ideal Fermi gas are strongly determined by the Pauli principle. We shall consider the limit:

$$k_BT \ll \mu, \qquad \beta\mu \gg 1,$$

which defines the degenerate Fermi gas. In this limit, the quantum mechanical nature of the system becomes especially important, and the system has little to do with the classical ideal gas.

Equation of State

Consider a gas of N non-interacting fermions, e.g., electrons, whose one-particle wave-functions $\varphi_r(\vec{r})$ are plane-waves. In this case, a complete set of quantum numbers r is given, for instance, by the three cartesian components of the wave vector $\vec{k}$ and the z spin projection m_s of an electron:

$$r \equiv \left(k_x,\ k_y,\ k_z,\ m_s\right).$$

Spin-independent Hamiltonians

We will consider only spin independent Hamiltonian operator of the type:

$$\hat{H} = \sum_k \epsilon_k\ c_k^\dagger c_k + \int d^3 r V(r)\ c_r^\dagger\ c_r,$$

where the first and the second terms are respectively the kinetic and th potential energy. The summation over the states r (whenever it has to be performed) can then be reduced to the summation over states with different wavevector $k\ (p = \hbar k)$:

$$\sum_r \ldots \quad \Rightarrow \quad (2s+1) \sum_k \ldots,$$

where the summation over the spin quantum number $m_s = -s, -s+1,\ \ldots,\ s$ has been taken into account by the prefactor (2s + 1).

Wave-functions in a Box

We assume that the electrons are in a volume defined by a cube with sides $L_x,\ L_y,\ L_z$ and volume $V = L_x L_y L_z$. For the one-particle wave function:

$$\langle r \mid k \rangle = \psi_k(r) = \frac{1}{\sqrt{V}}\ e^{ik \cdot r}$$

We use periodicity condition, here along the x-direction, at the cube's walls:

$$e^{ik_x x} = e^{ik_x x + ik_x L_x},$$

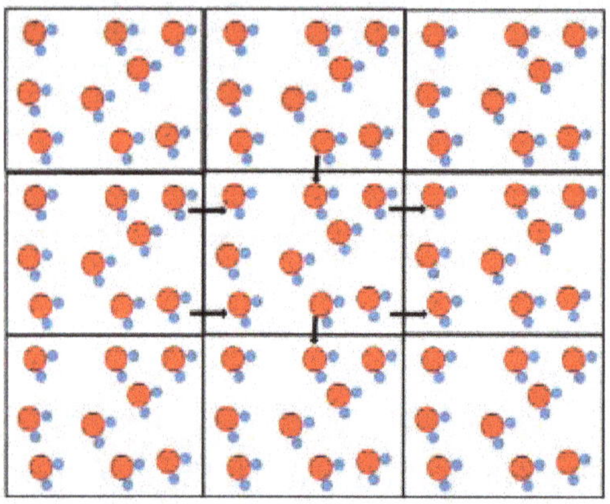

Which is then translated into a condition for the allowed k-values:

$$e^{ik_xL_x} = e^{i2\pi n_x}, \qquad k_x = \frac{2\pi}{L_x} n_x, \qquad n_x \in Z.$$

Analogously for the y- and for the z direction.

Summation over Wavevectors

Each state has in k-space an average volume of:

$$\Delta k = \frac{(2\pi)^3}{L_x L_y L_z} = \frac{(2\pi)^3}{V}.$$

For large $V \to \infty$ we can then replace the sum Σ_r over all quantum number by:

$$\sum_r \to (2s+1)\ \frac{1}{\Delta k}\ \int d^3k = (2s+1)\ \frac{V}{(2\pi)^3} \int d^3k$$
$$= (2s+1)\frac{V}{h^3}\int d^3p,$$

Where, $k = p/\hbar$ has been used.

Grand Canonical Potential

We consider now the expression for the fermionic grand canonical potential $\Omega(T, V, \mu)$,

$$\Omega(T, V, \mu) = -k_B T\ \sum_r \ln\left[1 + e^{-\beta(\epsilon_r - \mu)}\right].$$

Using the substitution $\sum_r \to (2s+1)\ \frac{1}{\Delta k}\ \int d^3k = (2s+1)\ \frac{V}{(2\pi)^3}\int d^3k = (2s+1)\frac{V}{h^3}\int d^3p$, and $\int d^3k = 4\pi\int k^2 dk$ we write the grand canonical potential as,

$$-\beta\,\Omega(T, V, \mu) = (2s+1)\frac{V}{(2\pi)^3}\,4\pi\int_0^\infty dk\,k^2\,\ln\left[1 + z\,e^{-\beta\hbar^2k^2/(2m)}\right],$$

where we used the usual expressions:

$$z = e^{\beta\mu}, \qquad \epsilon_r \to \epsilon_k = \frac{\hbar^2k^2}{2m}$$

For the fugacity z and for the one-particle dispersion and used an explicit expression for the one-particle energies for free electrons ϵ_k.

1. Dimensionless variables:

Expression $-\beta\,\Omega(T,\,V,\,\mu)=(2s+1)\dfrac{V}{(2\pi)^3}4\pi\int_0^\infty dk\,k^2\ln\left[1\;+\;z\;e^{-\beta\hbar^2k^2/(2m)}\right]$, is transformed further by introducing with:

$$x=\hbar k\sqrt{\frac{\beta}{2m}},\qquad k^2dk=\left(\frac{2m}{\beta\hbar^2}\right)^{3/2}x^2\,dx$$

A dimensionless variable x. One obtains:

$$-\beta\,\Omega(T,\,V,\,\mu)=(2s+1)\frac{4V}{\sqrt{\pi}}\left(\frac{m}{2\pi\beta\hbar^2}\right)^{3/2}\int_0^\infty x^2\,dx\ln\left(1+z\,e^{-x^2}\right).$$

2. De Broglie wavelengths:

By making use of the definition of the thermal de Broglie wavelength λ,

$$\lambda\;=\;\sqrt{\frac{2\pi\beta\hbar^2}{m}},$$

we then get:

$$-\beta\,\Omega(T,\,V,\,\mu)\;=\frac{(2s+1)}{\lambda^3}\,\frac{4V}{\sqrt{\pi}}\int_0^\infty dx\,x^2\ln\left(1+z\,e^{-x^2}\right).$$

3. Term by term integration:

We use the Taylor expansion of the logarithm,

$$\ln(1+y)=\sum_{n=1}^{\infty}(-1)^{n+1}\frac{y^n}{n},$$

In order to evaluate the integral,

$$\begin{aligned}\int_0^\infty x^2\,dx\ln\left(1+z\,e^{-x^2}\right)\;&=\;\sum_{n=1}^{\infty}(-1)^{n+1}\frac{z^n}{n}\int_0^\infty dx\,x^2\;e^{-nx^2}\\&=\sum_{n=1}^{\infty}(-1)^{n+1}\frac{z^n}{n}\left(-\frac{d}{dn}\int_0^\infty dx\,e^{-nx^2}\right)\\&=\;\sum_{n=1}^{\infty}(-1)^{n+1}\frac{z^n}{n}\left(-\frac{d}{dn}\,\frac{1}{2}\sqrt{\pi}\,\frac{1}{\sqrt{n}}\right)\end{aligned}$$

term by term. The result is:

$$\int_0^\infty x^2\,dx\ln\left(1+z\,e^{-x^2}\right)=\frac{\sqrt{\pi}}{4}\sum_{n=1}^{\infty}(-1)^{n+1}\frac{z^n}{n^{5/2}}.$$

4. Grand canonical potential:

Defining:

$$f_{5/2}(z) = \sum_{n=1}^{\infty}(-1)^{n+1}\frac{z^n}{n^{5/2}} = \frac{4}{\sqrt{\pi}}\int_0^\infty dx\, x^2 \ln\left(1+z\, e^{-x^2}\right)$$

we obtain:

$$\beta\,\Omega(T,\,V,\,\mu) = -\frac{2s+1}{\lambda^3}\,V f_{5/2}(z)$$

For the grand canonical potential for an ideal Fermi gas.

5. Pressure:

Our result $\beta\,\Omega(T,\,V,\,\mu) = -\frac{2s+1}{\lambda^3}\,V f_{5/2}(z)$ reduces with $\Omega = -PV$ to:

$$\frac{P}{k_BT} = \frac{2s+1}{\lambda^3} f_{5/2}(z), \qquad \lambda = \sqrt{\frac{2\pi\hbar^2}{k_BTm}}, \qquad z = e^{\mu/(k_BT)},$$

which yields the pressure $P = P(T,\,\mu)$.

6. Density:

With:

$$\Omega = -k_BT\,\ln Z = -PV \qquad \frac{PV}{k_BT} = \ln Z$$

We find:

$$\langle\hat{N}\rangle = z\left(\frac{\partial}{\partial z}\ln Z\right)_{T,V} = V\,z\left(\frac{\partial}{\partial z}\frac{P}{k_BT}\right)_{T,V}$$

For the number of particles N. The density $n(T,\,\mu) = \langle\hat{N}\rangle / V$ is then given by:

$$n = \frac{\langle\hat{N}\rangle}{V} = \frac{2s+1}{\lambda^3(T)} f_{3/2}(z).$$

where we have defined:

$$f_{3/2}(z) = z\frac{d}{dz}f_{5/2}(z) = \sum_{n=1}^{\infty}(-1)^{n+1}\frac{z^n}{n^{3/2}}.$$

Thermal Equation of State

As a matter of principle one could solve $n = \frac{\langle \hat{N} \rangle}{V} = \frac{2s+1}{\lambda^3(T)} f_{3/2}(z)$. for the fugacity z = z(T, n), which could then be used to substitute the fugacity in $\frac{P}{k_B T} = \frac{2s+1}{\lambda^3} f_{5/2}(z),\ \lambda = \sqrt{\frac{2\pi\hbar^2}{k_B T m}},\ z = e^{\mu/(k_B T)}$ for T and n, yielding such the thermal equation of state. This procedure can however not be performed in closed form.

Rewriting the particle density:

The function $f_{3/2}(z)$ entering the expression $n = \frac{\langle \hat{N} \rangle}{V} = \frac{2s+1}{\lambda^3(T)} f_{3/2}(z)$. for the particle density may be cast into a different form. Helping is here the result $\frac{P}{k_B T} = \frac{2s+1}{\lambda^3} f_{5/2}(z),\ \lambda = \sqrt{\frac{2\pi\hbar^2}{k_B T m}},\ z = e^{\mu/(k_B T)}$ for P(T, μ):

$$\frac{P}{k_B T} = \frac{2s+1}{\lambda^3} f_{5/2}(z), \quad f_{5/2}(z) = \frac{\lambda^3}{2s+1}\frac{\ln \mathcal{Z}}{V}, \quad \frac{PV}{k_B T} = \ln \mathcal{Z},$$

which leads to:

$$f_{3/2}(z) = z\frac{d}{dz} f_{5/2}(z) = z\frac{d}{dz}\left(\frac{\lambda^3}{2s+1}\ \frac{\ln \mathcal{Z}}{V}\right)$$

With:

$$\frac{d}{dz} = \frac{d}{d\beta}\left(\frac{d\beta}{dz}\right) = \frac{1}{\mu z}\frac{d}{d\beta}, \qquad \beta = \frac{\ln z}{\mu}$$

And $\lambda = \sqrt{(2\pi\beta\hbar^2)/m}$ we then find:

$$\mu V(2s+1) f_{3/2}(z) = \frac{d}{d\beta}\left(\lambda^3 \ln \mathcal{Z}\ \right) = \frac{3\lambda^3}{2\beta}\ln \mathcal{Z} + \lambda^3\ \frac{d}{d\beta}\ln \mathcal{Z}.$$

For the particle density $n = \frac{\langle \hat{N} \rangle}{V} = \frac{2s+1}{\lambda^3(T)} f_{3/2}(z)$. we finally obtain:

$$n = \frac{2s+1}{\lambda^3(T)} f_{3/2}(z), \quad \mu n V = \frac{3}{2}PV + \frac{d}{d\beta}\ln \mathcal{Z}, \quad nV = N,$$

where we have used $\ln \mathcal{Z}/\beta = PV$.

Caloric Equation of State

The expression for $\mu nV = \mu N$ in $n = \frac{2s+1}{\lambda^3(T)} f_{3/2}(z)$, $\mu nV = \frac{3}{2}PV + \frac{d}{d\beta}\ln \mathcal{Z}$, $nV = N$, leads with:

$$U = -\frac{d}{d\beta}\ln \mathcal{Z} + \mu\langle \hat{N}\rangle, \quad \mathcal{Z} = \sum_r e^{-\beta(\epsilon_r - \mu)}$$

to the caloric equation of state:

$$U = \frac{3}{2}PV$$

The equation U = 3PV/2 is also valid for the classical ideal gas, but it is not anymore valid for relativistic fermions.

Classical Limit

Starting from the general formulas for P(T, μ) and n(T, μ), we first investigate the classical limit (i.e. the non-degenerate Fermi gas), which corresponds:

$$n\lambda^3 \ll 1, \quad z = e^{\beta\mu} \ll 1.$$

Under this condition, the Fermi-Dirac distribution function reduces to the Maxwell Boltzmann distribution function:

$$\langle \hat{n}_r\rangle = \frac{1}{z^{-1}e^{\beta\epsilon_r} + 1} \approx ze^{-\beta\epsilon_r}$$

Expansion in the Fugacity

For a small fugacity z we may retain in the series expansion for $f_{5/2}(z)$ and $f_{3/2}(z)$, compare

$$f_{5/2}(z) = \sum_{n=1}^{\infty}(-1)^{n+1}\frac{z^n}{n^{5/2}} = \frac{4}{\sqrt{\pi}}\int_0^{\infty} dx\, x^2 \ln\left(1 + z\, e^{-x^2}\right) \quad \text{and} \quad f_{3/2}(z) = z\frac{d}{dz}f_{5/2}(z) = \sum_{n=1}^{\infty}(-1)^{n+1}\frac{z^n}{n^{3/2}},$$

the first terms:

$$f_{5/2}(z) \approx z - \frac{z^2}{2^{5/2}} \qquad \beta P\lambda^3 \approx (2s+1)z\left(1 - \frac{z}{2^{5/2}}\right)$$

$$f_{3/2}(z) \approx z - \frac{z^2}{2^{3/2}} \qquad n\lambda^3 \approx (2s+1)z\left(1 - \frac{z}{2^{3/2}}\right)$$

High-temperature Limit

The expression for $n\lambda^3$ in above equation reduces in lowest approximation to:

$$z^{(0)} \approx \frac{n\lambda^3}{2s+1}, \quad n\lambda^3 \ll 1.$$

The number of particles nλ³ in the volume spanned by the Broglie wavelength $\lambda \sim 1/\sqrt{T}$ is hence small are small. This is the case at elevated temperatures.

Fugacity Expansion

Expanding in the fugacity $z = exp\left(\beta\mu\right)$

$$z \approx \underbrace{\frac{n\lambda^3}{2s+1}}_{z^{(0)}} \frac{1}{1 - z2^{-3/2}} \approx \frac{z(0)}{1 - z^{(0)}2^{-3/2}}, \qquad z^{(1)} \approx z^{(0)} \left(1 + z^{(0)}2^{-3/2}\right),$$

where $1/\left(1-x\right) \approx 1 + x$ for $x \ll 1$ was used. The equation of state for the pressure, namely βPλ³ ≈ (2s + 1)(z – z22−5/2, is then:

$$\beta P\lambda^3 \approx \left(2s+1\right) \left[z^{(0)} \; (1 + z^{(0)}2^{-3/2}) - 2^{-5/2} \, (z^{(0)})^2\right]$$

$$= n\lambda^3 \left(1 \; + \; 2^{-5/2} \; \frac{n\lambda^3}{2s+1}\right),$$

when:

$$\frac{1}{\sqrt{2}^3} - \frac{1}{\sqrt{2}^5} = \frac{2}{\sqrt{2}^5} - \frac{1}{\sqrt{2}^5} = \frac{1}{\sqrt{2}^5}$$

is used. Altogether we then find:

$$PV = \left\langle \hat{N} \right\rangle k_B T \left(\; 1 \; + \; \frac{n\lambda^3}{4\sqrt{2}\left(2s+1\right)} \; \right).$$

In this expression, the first term corresponds to the equation of state for the classical ideal gas, while the second term is the first quantum mechanical correction.

Degenerated Fermi Gas

In the low temperature limit, $T \rightarrow 0$, the Fermi distribution function behaves like a step function:

$$n_k = \frac{1}{e^{\beta(\epsilon_k - \mu)} + 1} \xrightarrow{T \rightarrow 0} \begin{cases} 0 & \text{if } \epsilon_k > \mu \\ 1 & \text{if } \epsilon_k < \mu \end{cases}$$

i.e.,

$$\lim_{T \rightarrow 0} n_k = \theta\left(\mu \; - \; \epsilon_k\right).$$

Fermi Energy

This means that all the states with energy below the Fermi energy $\in_F$,

$$\in_F = \mu(n,\ T = 0),$$

are occupied and all those above are empty.

Fermi Sphere

In momentum space the occupied states lie within the Fermi sphere of radius p_F. The system is then deep in the quantum regime.

The Fermi energy is then be determined by the condition that the the Fermi sphere contains the correct number of states:

$$N = \sum_{\text{states r with } \in_r < \in_F} 1,$$

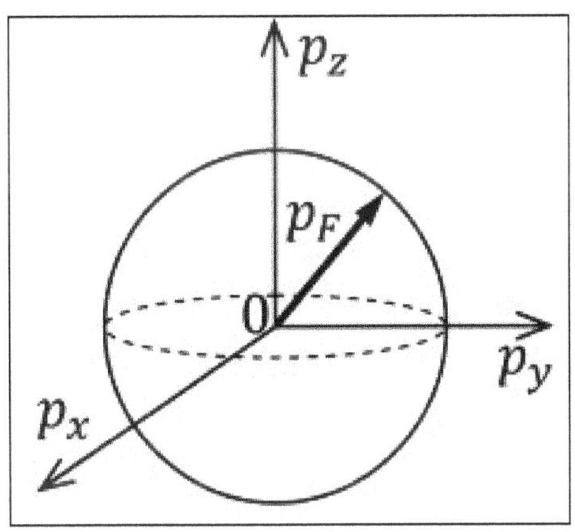

Which can be written for the case of free fermions:

$$N = \frac{(2s+1)V}{(2\pi)^3}\int_{|k|<|k_F|} d^3k = \frac{(2s+1)V}{(2\pi)^3}\frac{4}{3}\pi k_F^3$$

Here, $k_F = \frac{p_F}{\hbar}$ is the Fermi wave number. We have:

$$n = \frac{N}{V} = \frac{2s+1}{6\pi^2}k_F^3, \quad k_F = \left(\frac{6\pi^2 n}{2s+1}\right)^{1/3}$$

The Fermi energy is then:

$$\in_F = \frac{\hbar^2 k_F^2}{2m}, \quad \in_F = \frac{\hbar^2}{2m}\left(\frac{6\pi^2 n}{2s+1}\right)^{2/3}$$

Ground State Properties

At T = 0, the system is in its ground state, with the internal energy U_0 given by:

$$U_0 = \sum_{|\vec{k}|<k_F} \in_F = \frac{(2s+1)V}{(2\pi)^3} \int_0^{k_F} dk \left(4\pi k^2\right) \frac{\hbar^2 k^2}{2m}$$
$$= \frac{(2s+1)V}{(2\pi)^3} \left(\frac{\hbar^2}{2m}\right) \frac{4\pi}{5} k_F^5$$

Using the expression for total particle number N,

$$N = \frac{V(2s+1)}{(2\pi)^3} \frac{4\pi}{3} k_F^3,$$

for $k_F^5 = k_F^2 k_F^2$, one obtains:

$$\frac{U_0}{N} = \frac{3}{5}\frac{\hbar^2 k_F^2}{2m}, \quad \frac{U_0}{N} = \frac{3}{5} \in_F \quad \text{(independent of } s\text{)}$$

For internal energy per particle at absolute zero.

Pressure

Since PV = 2U/3, we obtain now an expression for the zero-point pressure P_0:

$$P_0 \equiv P_{T=0} = \frac{2}{5} n \in_F$$

- The zero-point pressure arises from the fact that fermionic particles move even at absolute zero. This is because the zero-momentum state can hold only one particle of a given spin state.
- Taking a Fermi energy of typically $\in_F \approx 10\ eV = 16 \cdot 10^{-19}$ and an electron density of $n \approx 10^{22} \cdot 100^3\ \text{m}^{-3}$ we find a zero-point pressure of:

$$P_0 \approx 3.2 \cdot 10^3 \cdot 10^6 \frac{J}{\text{m}^3} \approx 3.2 \cdot 10^4\ \text{bar},$$

where we have used that $1\,P = 1\,J/\text{m}^3 = 10^{-5}$ bar.

Fermi Temperature

At low but a finite temperature, the Fermi distribution function $\langle \hat{n} \rangle\ \ n(\in)$ for the occupation number smooths out around the Fermi energy.

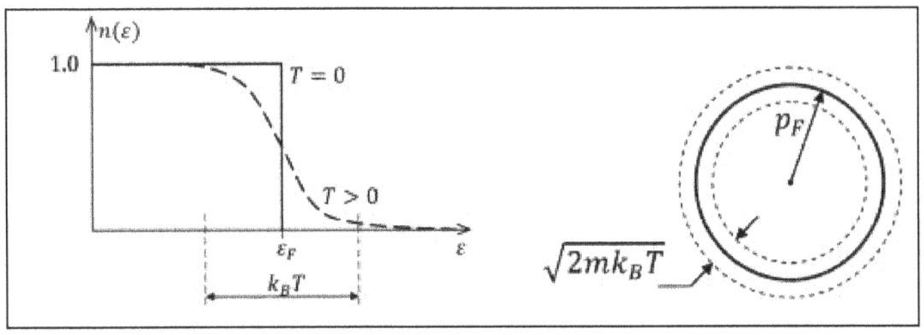

Such an evolution of $n(\in)$ with increasing temperature is due to the excitation of fermions within a layer beneath the Fermi surface to a layer above. "Holes" are left beneath the Fermi surface.

Fermi Temperature

We define the Fermi temperature T_F as:

$$\in_F = k_B T_F .$$

a) $T \ll T_F$

For low temperatures $T \ll T_F$, the Fermi distribution deviates from that at T = 0 mainly in the neighborhood of of in a layer of thickness $k_B T$. Particles at energies of order $k_B T$ below the Fermi energy are excited to energies of order $k_B T$ above the Fermi energy.

b) $T \gg T_F$

For $T \gg T_F$, the Fermi distribution approaches the Maxwell-Boltzmann distribution. The quantum nature of the constituent particles becomes irrelevant.

Frozen vs. Active Electrons

The typical magnitude,

$$\in_F \approx 2\ eV, \quad T_F \approx 2 \times 10^4\ K,$$

of the Fermi temperature in metals implies that room temperature electrons are frozen mostly below the Fermi level. Only a fraction of the order of:

$$\frac{T}{T_F} \approx 0.015$$

of the electrons contributes to thermodynamic properties involving excited states.

1. Particles and holes

We can define that the absence of a fermion of energy $\in$, momentum $\vec{p}$ and charge e corresponds to the presence of a hole with:

$$\begin{aligned} \text{energy} &= -\in \\ \text{momentum} &= -\vec{p} \\ \text{charge} &= -e \end{aligned}$$

The concept of a hole is useful only at low temperatures $T \ll T_F$, when there are few holes below the Fermi surface. The Fermi surface "disappears" when $T \gg T_F$, with the system approaching the Maxwell-Boltzmann distribution function.

2. Specific heat

Since the average excitation energy per particle is $k_B T$, the internal energy of the system is of order:

$$U \approx U_0 + \left(\frac{T}{T_F}\right) N k_B T,$$

where U_0 is the ground-state energy. The specific heat capacity C_V is then of the order of:

$$\frac{C_V}{Nk_B} \sim \frac{T}{T_F}, \quad C_V = \left(\frac{\partial U}{\partial T}\right)_V$$

The electronic contribution to the specific heat vanishes linearly with $T \to 0$. The roomtemperature contribution of phonons (lattice vibrations) to C_V is therefore in general dominant.

Low Temperature Expansion

Density of States

We will work from now on with density of state per volume $D(E) = \Omega(E)/V$, which is defined as the derivative of the integrated phase space per volume, $\varphi(E) = \Phi(E)/V$:

$$D(E) = \frac{\partial \varphi(E)}{\partial E}, \quad \varphi(E) = \frac{1}{V}\int_{\epsilon_k \le E} \frac{d^3k}{\Delta k^3}, \quad \Delta k^3 = \frac{(2\pi)^3}{V}.$$

The spin degeneracy factor 2s + 1 will be added further below.

1. Fermi sphere

With the phase space being isotropic we may write the volume of the Fermi sphere as,

$$\left(V\Delta k^3\right)\varphi(E) = \frac{4}{3}\pi k_E^3 = \frac{4\pi}{3}\left(\frac{2mE}{\hbar^2}\right)^{3/2}, \quad E = \frac{\hbar^2 k_E^2}{2m}.$$

Introducing the spin degeneracy factor 2s + 1 we then obtain:

$$D(E) = \begin{cases} A\sqrt{E} & \text{if } E \ge 0, \\ 0 & \text{otherwise,} \end{cases}$$

$$A = \frac{2s+1}{(2\pi)^2}\left(\frac{2m}{\hbar^2}\right)^{3/2}$$

Since we did not use any special properties of a fermionic system, this expression is also valid for bosons. For both types of systems, D(E) shows a $\sqrt{E}$ dependence.

4. Energy and particle density

The energy density U/V is given by:

$$\frac{U(T,\mu)}{V} = \int_{-\infty}^{+\infty} dE\; E\; n(E)D(E), \qquad n(E) = \frac{1}{e^{\beta(E-\mu)}+1},$$

where n(E) is the Fermi distribution as a function of the energy. The analogous expression for the particle density is:

$$n(T,\mu) = \frac{\langle \hat{N} \rangle}{V} = \int_{-\infty}^{+\infty} dE\; n(E)D(E)$$

For a bosonic system one substitutes the Boson distribution function $1/\left(e^{\beta(E-\mu)}-1\right)$ for n(E).

Sommerfeld Expansion

We are interested in the thermodynamic properties of a fermionic system at small but finite temperature, viz at a low temperature expansion of expectation values like above equations.

We denote with *H(E)* a function depending exclusively on the one-particle energy *E*. We will show that:

$$\langle H \rangle = \int_{-\infty}^{\infty} dEH(E)n(E) \approx \int_{-\infty}^{\mu} dEH(E) + \frac{\pi^2}{6}(k_BT)^2\, H'(\mu)$$

Holds terms or order T^4 or higher.

- The first term on the r.h.s. of above equaton survives when $T \rightarrow 0$ and $n(E) \rightarrow \theta(\mu - E)$. It represents the ground-state expectation value.
- The second term results from expanding both H(E) and n(E) around the chemical potential μ.

Partial Integration

We start the derivation of the Sommerfeld expansion with the definition of partial integration.

$$K(E) = \int_{-\infty}^{E} dE'\, H(E'), \quad H(E) = \frac{dK(E)}{dE},$$

which allows as to perform the partial integration:

$$\langle H \rangle = \int_{-\infty}^{+\infty} dE\, \frac{dK(E)}{dE}\, n(E) = \int_{-\infty}^{+\infty} dE\, K(E) \left(\frac{-dn(E)}{dE}\right).$$

1. For the integration we have used $\lim_{E\to\infty} n(E) = 0$, namely that the probability to find particles at elevated energies E falls of exponentially.

2. We have also assumed that $\lim_{E\to-\infty} H(E) = 0$.

Taylor Expansion

Substituting the first two terms of the Taylor expansion:

$$K(E)=K(\mu)+(E-\mu)K'(\mu)+\frac{(E-\mu)^2}{2}K''(\mu)+O\left((E-\mu)^3\right)$$

of K(E) around the chemical potential μ into previous equation leads to:

$$\langle H\rangle \approx \int_{-\infty}^{+\infty} dE\left[K(\mu)+(E-\mu)K'(\mu)+\frac{(E-\mu)^2}{2}K''(\mu)\right]\left(\frac{-dn(E)}{dE}\right)$$

Ground State Contribution

The first term in previous equation is:

$$K(\mu)\int_{-\infty}^{\infty} dE\left(-\frac{dn(E)}{dE}\right)=K(\mu)\left[n(-\infty)-n(\infty)\right] = \int_{-\infty}^{\mu} dEH(E),$$

viz the $T=0$ value of $\langle H\rangle$.

Symmetry Cancellation

The second term in $\langle H\rangle \approx \int_{-\infty}^{+\infty} dE\left[K(\mu)+(E-\mu)K'(\mu)+\frac{(E-\mu)^2}{2}K''(\mu)\right]\left(\frac{-dn(E)}{dE}\right)$ vanishes because:

$$-\frac{dn}{dE}=\frac{\beta e^{\beta(E-\mu)}}{\left[e^{\beta(E-\mu)}+1\right]^2}=\frac{\beta e^{-\beta(E-\mu)}}{\left[1+e^{-\beta(E-\mu)}\right]^2}, \qquad n(E)=\frac{1}{e^{\beta(E-\mu)}+1},$$

is symmetric in E – μ.

Finite Temperature Correction

The third term in $\langle H\rangle \approx \int_{-\infty}^{+\infty} dE\left[K(\mu)+(E-\mu)K'(\mu)+\frac{(E-\mu)^2}{2}K''(\mu)\right]\left(\frac{-dn(E)}{dE}\right)$

yields the first non-trivial correction:

$$K''(\mu)\int_{-\infty}^{+\infty} dE\frac{(E-\mu)^2}{2}\left(-\frac{dn(E)}{dE}\right) = \frac{\pi^2}{6}(k_BT)^2 H'(\mu),$$

where the scaling with $(k_B T)^2 = 1/\beta^2$ follows from a transformation to dimensionless variables $y = \beta(E - \mu)$.

The factor $\frac{\pi^2}{6}$ results from the final dimensionless integral. This concludes our derivation of the Sommerfeld expansion $\langle H \rangle = \int_{-\infty}^{\infty} dEH(E)n(E) \approx \int_{-\infty}^{\mu} dEH(E) + \frac{\pi^2}{6}(k_B T)^2 H'(\mu)$.

Internal Energy at Low Temperatures

We start applying the Sommerfeld expansion:

$\langle H \rangle = \int_{-\infty}^{\infty} dEH(E)n(E) \approx \int_{-\infty}^{\mu} dEH(E) + \frac{\pi^2}{6}(k_B T)^2 H'(\mu)$ to the particle density

$$n(T, \mu) = \frac{\langle \hat{N} \rangle}{V} = \int_{-\infty}^{+\infty} dE\, n(E)D(E)$$

$$\begin{aligned} n(T, \mu) &= \int_{-\infty}^{+\infty} dE\, D(E)n(E) \\ &= \int_{-\infty}^{\mu} dE\, D(E) + \frac{\pi^2}{6}(k_B T)^2 D'(\mu) \\ &= \int_{-\infty}^{\epsilon_F} dE\, D(E) + \left\{ \underbrace{\int_{\epsilon_F}^{\mu} dE\, D(E)}_{\approx (\mu - \epsilon_F) D(\epsilon_F)} + \frac{\pi^2}{6}(k_B T)^2 D'(\mu) \right\}, \end{aligned}$$

where we have taken into account that $\mu = \mu(T)$ may be different from the Fermi energy $\epsilon_F = \lim_{T \to 0} \mu(T)$. We have also assumed that the density of states D(E) is essentially constant around the Fermi energy.

Constant Particle Density

The terms inside the bracket in previous equation need to cancel if the particle density n is to be constant. The chemical potential μ varies hence as,

$$\mu = \epsilon_F - \frac{\pi^2}{6}(k_B T)^2 \frac{D'(\epsilon_F)}{D(\epsilon_F)}, \quad \mu = \epsilon_F - \frac{\pi^2}{12} \frac{(k_B T)^2}{\epsilon_F}$$

as a function of temperature.

1. We have approximated $D'(\mu)$ in previous equation consistently by $D'(\epsilon_F)$.
2. The scaling $D(E) \sim \sqrt{E}$, as given by $D(E) = \begin{cases} A\sqrt{E} & \text{if } E \geq 0, \\ 0 & \text{otherwise,} \end{cases} \quad A = \frac{2s+1}{(2\pi)^2}\left(\frac{2m}{\hbar^2}\right)^{3/2}$, leads to $D'/D = 1/(2E)$.

Internal Energy

The energy density $\frac{U(T,\mu)}{V} = \int_{-\infty}^{+\infty} dE\ E\ n(E)D(E), \quad n(E) = \frac{1}{e^{\beta(E-\mu)}+1}$ evaluated with the Sommerfeld expansion is:

$$u = U/V = \int_{-\infty}^{+\infty} dE\ E\ n(E)D(E)$$

$$= u_0 + \int_{\epsilon_F}^{\mu} dE\ E\ D(E) + \frac{\pi^2}{6}(k_BT)^2 \underbrace{\frac{d}{dE}(ED(E))\Big|_{E=\mu}}_{\frac{3}{2}D(\epsilon_F)},$$

where $u_0 = \int_{-\infty}^{\epsilon_0} ED(E)dE$ is the ground state energy. The substitution $\mu \to \epsilon_F$ performed for the argument of last term, for which we used $D(E) \sim \sqrt{E}$, is correct to order T². The second term is:

$$\int_{\epsilon_F}^{\mu} dE\ E\ D(E) \approx (\ \mu - \epsilon_F\ )_{\epsilon_F}\ D_{(\epsilon_F)} = -\frac{\pi^2 (k_BT)^2}{12} D_{(\epsilon_F)}.$$

when $D(E) \sim \sqrt{E}$. One finds hence with $3/12 - 1/12 = 1/6$ that:

$$\frac{U}{V} = u_0 + \frac{\pi^2 (k_BT)^2}{6} D_{(\epsilon_F)} \quad u - u_0 \propto (k_BT)^2\ D_0.$$

The internal energy increases quadratically with the temperature, being at the same time proportional to the density of states $D_0 = D_{(\epsilon_0)}$ at the Fermi level.

- Specific heat:

Assuming a constant density of states $D(E) \approx D(\epsilon_F)$ close to the Fermi energy ϵ_F we find:

$$c_V = \frac{C_V}{V} = \frac{1}{V}\frac{\partial U}{\partial T}, \quad c_V \approx \frac{\pi^2}{3}(k_B^2 T)\ D(\epsilon_F)$$

for the intensive specific heat. A trademark of a fermionic gas is that c_V is linear in the temperature.

The heat capacity per volume saturates however for $T \to \infty$, where it becomes identical with the ideal gas value $C_V / V = 3nk_B / 2$.

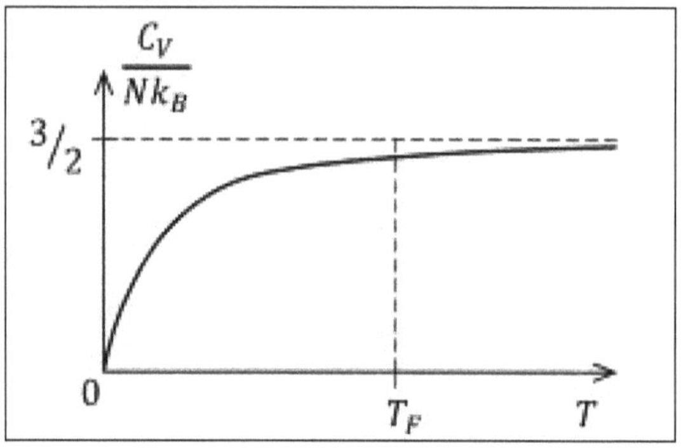

Gas Laws

All gases generally show similar behaviour when the conditions are normal. But with a slight change in physical conditions like pressure, temperature or volume these show a deviation. Gas laws are an analysis of this behaviour of gases. The variables of state like the Pressure, Volume and Temperature of a gas depict its true nature. hence gas laws are relations between these variables.

Boyle's Law

Boyle's Law is a very important gas law in chemistry and physics. It is also known as Mariotte's law or the Boyle-Mariotte law. The law is the oldest gas law. Boyle's law along with Charles's law, Gay-Lussac's law, and Avogadro's law forms the ideal gas law. The law correlates how the pressure of a gas increases with a decrease in the volume of the gas.

Boyle's law states for a fixed amount of an enclosed ideal gas, its absolute pressure is inversely proportional to its volume at a constant temperature.

Formula: The equation of Boyle's law is: $\mathrm{PV} = \text{constant}$.

As the law states pressure and volume are inversely proportional at a constant temperature for a given mass of an ideal gas. The statement can mathematically be expressed as:

$$P \propto \frac{1}{V}$$

The above expression can be rearranged as:

$$PV = k$$

where k is a constant of proportionality.

The above expression is valid for a given mass of an ideal gas and at a constant temperature. Hence as the volume increases, the pressure of the gas decreases and as the volume decreases, the pressure increases.

If P_1 and P_2 are the pressures at volumes V_1 and V_2 respectively at a constant temperature for a given amount of an ideal gas, then from Boyle's law:

$$P_1V_1 = P_2V_2$$

From the above expression, it is clear that the product of the pressure and the volume of an ideal gas is constant at a constant temperature. When the pressure of a gas at a constant temperature is double, the volume reduces to half of the initial volume. This is explained as below:

$$P_2 = 2P_1$$

$$P_1V_1 = 2P_1V_2$$

$$V_2 = \frac{V_1}{2}$$

Graphical Representation of Boyle's Law

The graphical representation of Boyle's law can be demonstrated in the three graphs which are explained.

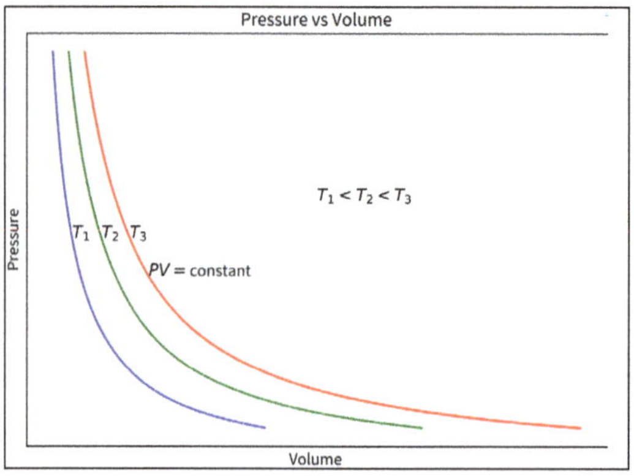

Figure: Pressure vs Volume Curves at Different Temperatures.

As can be observed from the above graph, the pressure of an ideal gas decreases with an increase in the volume. The above curve is hyperbolic in nature. It can also be observed that the curve shifts up with an increase in temperature. Each curve is at a constant temperature, which is called an isotherm.

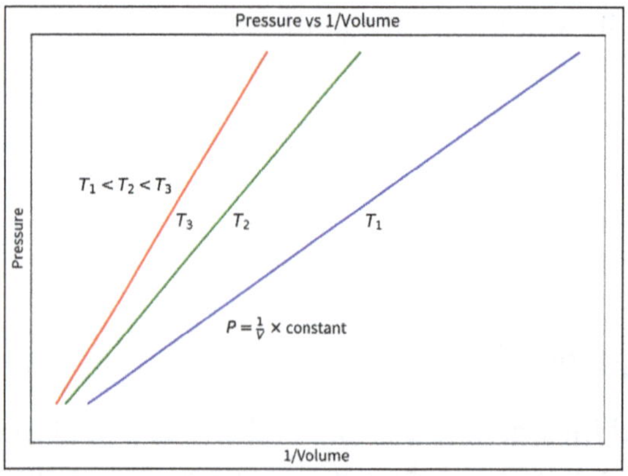

Figure: Pressure vs 1/Volume at Different Temperatures.

The graph of the pressure vs the inverse of the volume is a straight line passing through the origin. The slope of the line is the proportionality constant, *k*.

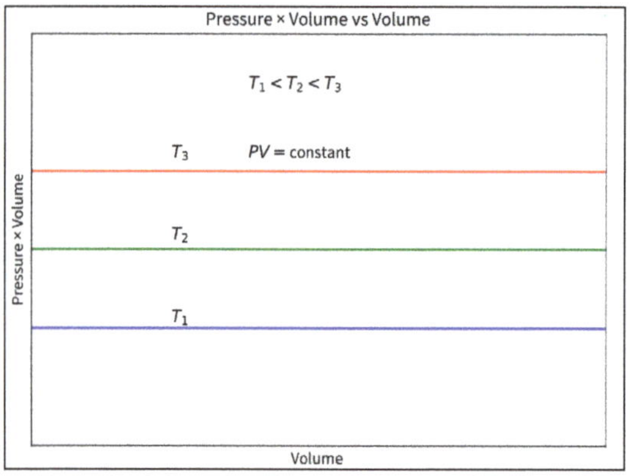

Figure: Pressure-Volume vs Volume at Different Temperatures.

The product of the pressure and the volume is always constant at a constant temperature for an ideal gas. This is observed from the above figure. Therefore, the product of the pressure and the volume is independent of the volume (or the pressure) at a constant temperature for an ideal gas.

We can also plot the logarithmic graphs of Boyle's law equation. The equation is as followed:

$$PV = k$$

Taking logarithm on both sides,

$$\log PV = \log k$$
$$\log P + \log V = \log k$$
$$\log P = -\log V + \log k$$
$$\log P = \log \frac{1}{V} + \log k$$

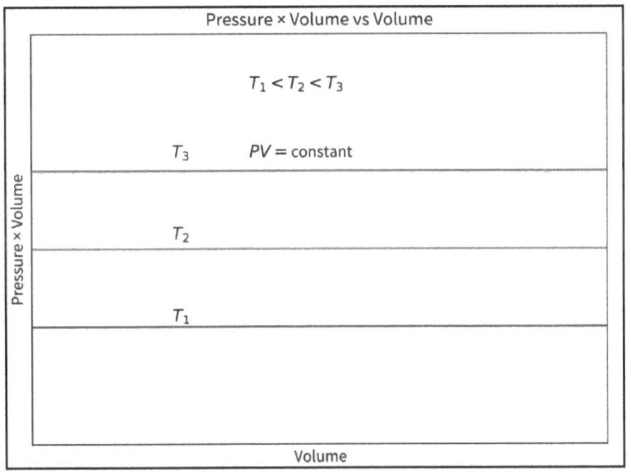

Figure: Logarithmic Graphs of Boyle's Law Equation (log(Pressure) vs log(Volume) on the left and log(Pressure) vs log(1/Volume) on the right).

Deriving Boyle's Law from Ideal Gas Law

We can also derive the Boyle's law from the ideal gas law. The famous ideal gas equation is mentioned hereunder.

$$PV = nRT$$

Where:

- n is the amount of gas (or moles of gas),
- R is the ideal gas constant, and
- T is the temperature.

Boyles' law has two primary conditions: constant temperature and a fixed amount of gas. Thus, T and n are constant. In the above ideal gas equation, the right-hand side of the equation, all the variables (n, R, and T) are constant. The product (nRT) of these three constant is also constant. So, we can replace nRT by the constant named k.

Finally, the equation becomes PV=*k*, which is nothing but Boyle's equation.

Limitation of Boyle's Law

- Boyle's law is applicable only to ideal gases.
- The law holds good only at high temperatures and low pressures.
- The law fails at high pressures. The product of the pressure and volume does not remain constant at high pressures but shows a slight increase. This increase is due to an increase in the volume which is caused by repulsive forces among the molecules. At high pressures, the molecules are too slow to one another. Repulsive forces become dominant as compactness increases. This results in the expansion of the gas.

Real-world Applications of Boyle's Law

There are numerous applications of Boyle's law. Some can be observed in day to day life. A few of them are mentioned below:

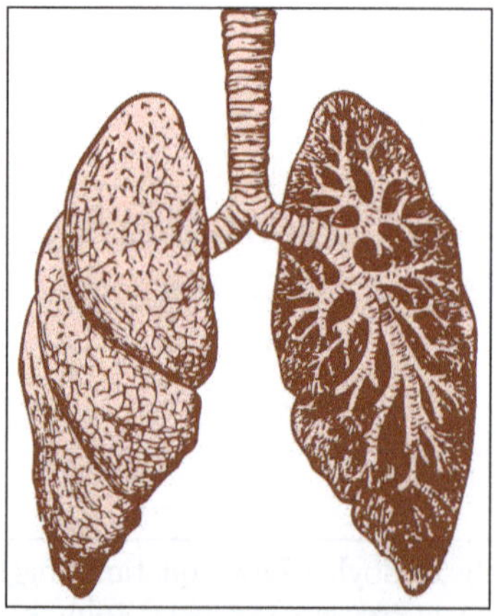

Figure: Human Lungs.

- Human Lungs: During the inhalation cycle of the breathing system, the lungs expand. This expansion is nothing but increase in the volume, which as per Boyle's law causes the reduction in the pressure relative to outside. Thus, air flows inside the body due to the pressure difference. By the same analogy, during the exhalation process, the volume decreases, which results in an increase in the pressure relative to the outside. And air flows from the body to outside.

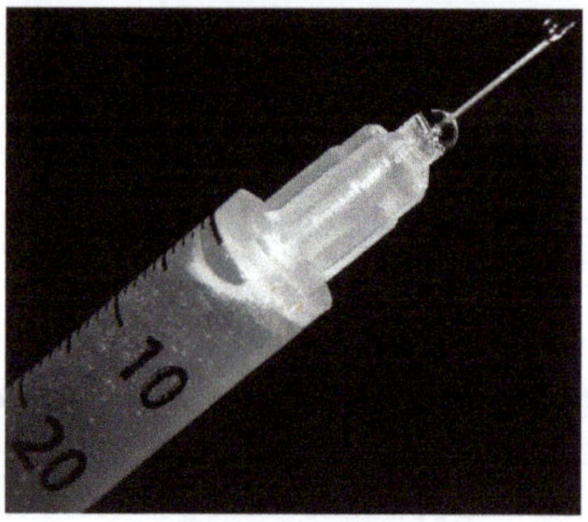

Figure: Syringe.

- Syringe: When the plunger of a syringe is pulled outside, there is the creation of vacuum i.e. a low-pressure region (due to increase in the volume) compare to outside and fluid gets suck in the syringe.

Figure: Soda Can with Opened Cap.

- Soda Cans: As we all know when the cap of a soda can is opened, the pressurised gas inside the can expands, which results in a decrease in pressure of the gas.

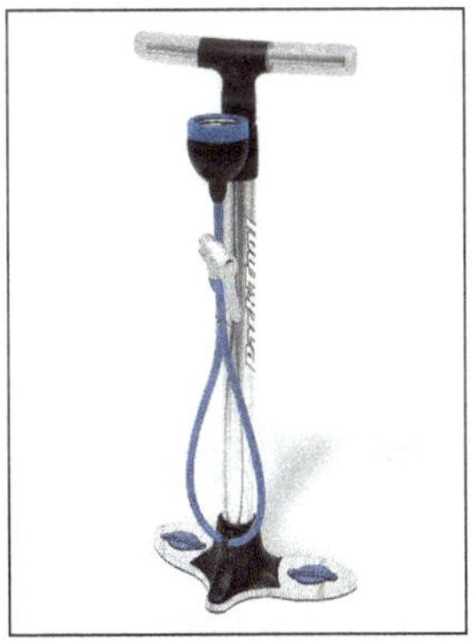

Figure: Hand Bicycle Pump.

- Hand Bicycle Pump: When the piston of a hand pump is pushed downwards, the air inside the pump compresses i.e. the volume reduces. This reduced volume increases the pressure and the air is pumped into tires.

Figure: Suba Diver.

- Scuba Divers: When a scuba diver dives beneath the water surface due to the hydrostatic pressure on divers, the air inside their lungs contracts. As a diver approaches the surface, the air inside their lungs expands since the pressure decreases on the surface of the water.

Example: To calculate expanded volume of methane.

Statement: Consider a fixed amount of methane gas at pressure $2\times10^{5}\,\mathrm{Nm}^{-2}$ of volume $2.0m^{3}$ goes under expansion such that the initial and final temperature remains the same. The new pressure of the gas is $1\times10^{5}\,\mathrm{Nm}^{-2}$. Calculate the new volume?

Solution: As from Boyle's law at a constant temperature and for a given mass of gas,

$$PV = k$$

$$P_1V_1 = P_2V_2$$

where P_1, V_1 and P_2, V_2 are the initial and final pressures and volumes.

$$V_2 = \frac{P_1V_1}{P_2} = \frac{2\times10^{5}\times2.0}{1\times10^{5}} = 4\mathrm{m}^{3}$$

Therefore, the new volume after the expansion is $4\mathrm{m}^3$.

Example: To Calculate New Pressure after Compression.

Statement: A storage tank A of volume 750 dm^3 contains a non-reactive gas at an absolute pressure of 4.0 bar. The gas is pumped into another storage tank B of volume 300 dm^3. Assume the temperatures of the gas inside both the tanks are the same, and there is no loss of the gas during the transportation. Calculate the new pressure of the gas?

Solution: Since the initial and final temperature are the same, and there is complete conservation of the mass of the gas during transportation, we can apply Boyle's law as:

$$P_AV_A = P_BV_B$$

Where P_A, V_A and P_B, V_B are the pressures and volumes of storage tank A and storage tank B respectively.

$$P_B = \frac{P_AV_A}{P_B} = \frac{4.0\times300}{750} = 1.6\,\mathrm{bar}$$

The pressure inside the storage tank B is 1.6 bar.

Example: To determine volume of hydrogen.

Statement: Hydrogen gas from four identical cylinders is charged inside a vessel. The absolute pressure inside each cylinder is 12 bar. The volume of each cylinder is 30 dm^3. The reading of the pressure gauge mounted on the vessel after charging of hydrogen gas is 3.3 bar g. Calculate the volume of the vessel using Boyle's law?

Solution: First, calculate the total initial volume which is the addition of all the volumes of the cylinders.

$$V_1 = 4\times30 = 120\ \mathrm{dm}^3$$

The final pressure is given in gauge, which needs to be converted into the absolute pressure. Absolute pressure is gauge pressure plus atmospheric pressure. Taking the atmospheric pressure as 1.0 bar,

$$P_2 = P_{gauge} + P_{atm} = 3.3 + 1.0 = 4.3\,\text{bar}$$

Assume the temperature remains constant in both systems. From Boyle's law,

$$P_1V_1 = P_2V_2$$

$$V_2 = \frac{P_1V_1}{P_2} = \frac{12\times120}{4.3} \approx 330\,\text{dm}^3$$

Therefore, the volume of the vessel (which the volume of gas occupied) is 330 dm^3.

Example: To determine volume of nitrogen.

Statement: For a given mass of nitrogen gas, the initial volume is 15 dm^3 at STP. The gas is expanded isothermally to the pressure of $5.45\times10^4 Nm^{-2}$. Calculate the change in the volumes of the gas?

Solution: At STP, the pressure is 1 atm. The initial pressure of the gas is,

$$P_1 = 1\,\text{atm} = 1.01325\times105\,\text{Nm}^{-2}$$

From Boyle's law,

$$V_2 = \frac{P_1V_1}{P_2} = \frac{15\times1.01325\times10^5}{5.45\times10^4}$$

$$= 27.9\,\text{dm}^3$$

Thus, the change in the volumes is calculated as:

$$\Delta V = V_2 - V_1 = 27.9 - 15 \approx 13\,\text{dm}^3$$

The gas is isothermally expanded by 13 dm^3.

Charle's Law

Charles's law is also known as the law of volumes. The law describes the relationship between the volume and temperature of a gas.

For a fixed amount of an enclosed ideal gas, its volume is directly proportional to its absolute temperature at a constant pressure.

Formula: $V = \text{constant} \times \text{T}$

As the law states: The volume and temperature of an ideal gas are directly proportional to each other at constant pressure and for a given mass of a gas. The statement can mathematically be expressed as:

$$V \propto T$$

By removing the proportionality, we get,

$$V = k \times T$$

where k is a constant of proportionality.

The above expression can be rearranged as:

$$\frac{V}{T} = k$$

The above expression is valid for a given mass of a gas and at a constant pressure. Hence, as the temperature increases, the gas expands and as the temperature decreases, the gas contracts. We can also conclude from the above equation, the ratio of the volume to absolute temperature is independent of temperature or volume at a constant pressure.

In the above expression temperature should always be on an absolute scale (in the kelvin).

For a given amount ideal gas, V_1, T_1 and V_2, T_2 are the volumes and temperatures of condition 1 and condition 2 at a constant pressure. From Charles's law,

$$\frac{V_1}{T_1} = \frac{V_2}{T_2}$$

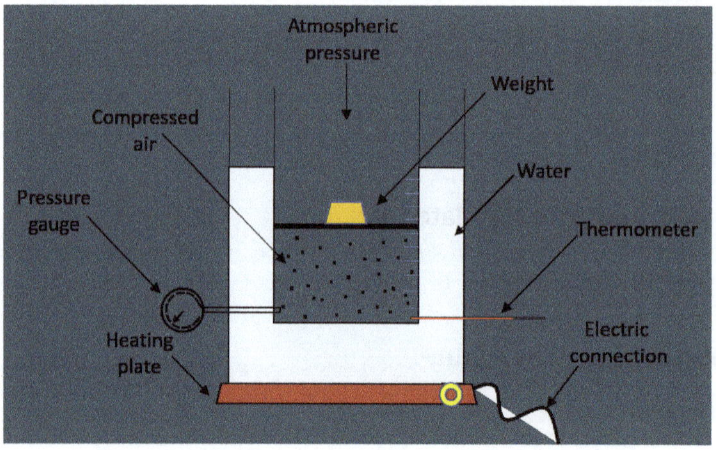

Figure: Experimental apparatus for Charles's law

The above diagram is a typical experimental setup required for verification of Charles's law. When the air inside the chamber is heated with a heating source (electric heater is used in the diagram) via water, the temperature of the air increases i.e. the kinetic energy associated with the molecules of the air increases. Thus, molecules of the air exert a force in the outwards direction on the resting plate and the air inside the chamber expands. By the same analogy, when the air is cooled, the volume of the air shrinks.

Alternative Version of Charles's Law

Charles from his experiments concluded that at constant pressure, the volume of a fixed amount of a gas increases or decreases by $\frac{1}{273}$ times the volume at $0\ ^{\circ}C$ for every $1\ ^{\circ}C$ rise or fall in temperature.

This statement can mathematically be expressed as:

$$V_T - V_0 = \frac{1}{273} \times V_0 \times t$$

where:

- V_T is the final expanded volume,
- V_0 is the volume at $0\ ^\circ C$,
- t is the temperature expressed in degree Celsius.

Rearranging the above expression,

$$V_T = \frac{1}{273} \times V_0 \times t + V_0 = V_0\left(1 + \frac{t}{273}\right)$$

$$V_T = V_0\left(\frac{273+t}{273}\right) = V_0\left(\frac{T}{273}\right) = V_0\left(\frac{T}{T_0}\right)$$

where T is the temperature in kelvin, and T_0 is the temperature in kelvin at $0\ ^\circ C$.

$$\frac{V_T}{V_0} = \frac{T}{T_0}$$

Thus, from the above equation, we arrive at Charles's law,

$$V \propto T$$

Charles's Law and Absolute Zero Temperature

As per Charles's law when the absolute temperature approaches zero, the volume of the gas should approach zero. According to Gay-Lussac, this could only happen if the gas is an ideal gas.

$$\lim_{T \to 0} V = 0$$

Graphical Representation

The graphical representation of Charles's law is demonstrated in the below graphs.

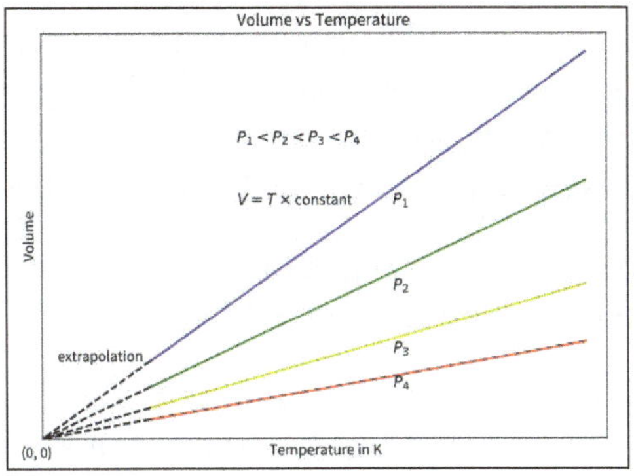

Figure: The graph of volume vs teperature (K) is a straight line passing through the origin.

From the graph, the volume increases linearly with an increase in the temperature. In the above figure, the temperature is taken in absolute scale. It can also be observed that as the temperature approaches zero, the volume also approaches zero. All the above four lines are plotted at a constant pressure i.e isobaric lines.

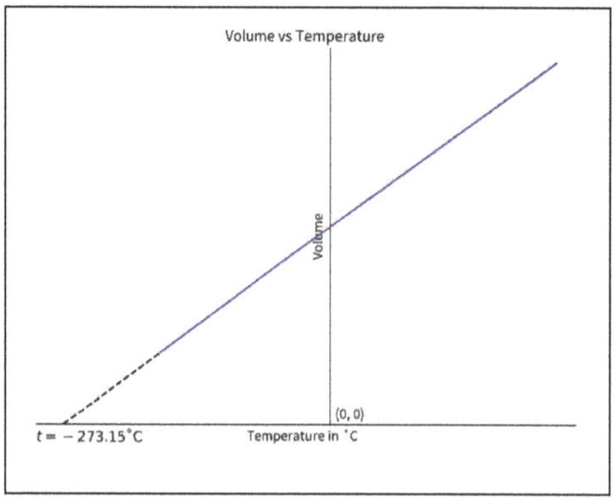

Figure: The graph of volume vs teperature (°C) is a straight line with x-intercept −273.15 °C.

The above line is plotted from using the below equation. The x-intercept in the above figure is $-273.15\,^{\circ}C$ and the y-intercept is V_0.

$$V = V + \frac{\quad}{273} t .$$

Real World Applications of Charles's Law

There are numerous applications of Charles's law can be observed in day to day life. Below are some of the mentioned:

1. Tyres: As we all have observed, sometimes, in the summer season the tyres of vehicles get busted. The reason is simple Charles's law. When a vehicle is left untouched for a prolonged period. As the summer season approaches, there is a rise in temperature which causes the air inside in tyres to expand. And after a certain threshold point, the tyres get ruptured. Therefore, a regular check is necessary to avoid any unaccepted consequences.

2. Hot air balloons: We all have seen hot air balloons flying. But has anyone wondered how it works? An Air Balloon consists of a bag, a basket to carry passengers, and a source of heat. The heat's source is usually a fuel like propane which on burning causes the heating of the air surrounding it. This as per Charles's law results in expansion of air i.e. the increase in volume. And with the increase in volume, the density of the air decreases. Hence, the buoyant force acts in upwards direction on the balloon and it flies.

3. Rubber balls: In the cold winter season, rubber balls get shrink while in hot summer, they swell. This shrinking and swelling are due to Charles's law.

4. Human lungs: Jogging is very difficult in the colder days to compare to hotter days. The reason is simply due to low temperatures the lungs shrink which decrease the human lungs capacity.

5. Ping Pong balls: In winter, usually the denting of Ping Pong balls occurs. The balls can be freed from this dent by applying heat gently. The heat warms up the air and it expands.

6. Pool floats: When inflated pool floats are pushed into pools, they appear as under-inflated. This is because the temperature of the pool is at a lower temperature than the surrounding atmosphere. Thus, pool floats shrink by Charles's law.

7. Bakery: Yeast is present in almost all bakery products. This yeast releases carbon dioxide gas. As the gas moves upwards, it is heated which causes its expansion. And making the cake or bread to rise and fluffy.

8. Spray Bottles: When spray bottles are placed into sunlight or any other heat source, the compressed air inside the bottles expands and sometimes can also burst the bottles. Hence, it is always recommended to keep the spray bottles away from the sunlight or high temperature.

Charles's law along with Boyle's law, Gay-Lussac's law and Avogadro's law form ideal gas law.

Examples of Charles's Law

1. Consider a fixed amount of propane gas at temperature $30\,^{\circ}\mathrm{C}$ of volume 1.5 m^3 goes under expansion such that the initial and final pressures remain the same. The new temperature of the gas is $75\,^{\circ}C$. Calculate the new volume.

First, convert both the temperatures in the problem statement from celsius to kelvin.

$$T_1 = 30 + 273.15 = 303.15\mathrm{K}$$
$$T_2 = 75 + 273.15 = 348.15K$$

As from Charles's law at constant pressure and for a given mass of gas,

$$\frac{V}{T} = k$$

$$\frac{V_1}{T_1} = \frac{V_2}{T_2}$$

$$V_2 = \frac{V_1}{T_1} \times T_2 = \frac{1.5}{303.15} \times 348.15 = 1.7\,\mathrm{m}^3$$

Therefore, the new volume after the expansion is 1.7 m^3.

2. A hot air balloon of volume 2.00 L is cooled from the temperature $65\,^{\circ}C$ to the room temperature. The final volume is 1.75 L. Estimate the room temperature?

The initial temperature T_1 is:

$$T_1 = 65 + 273.15 = 338.15K$$

From Charles's law,

$$\frac{V_1}{T_1} = \frac{V_2}{T_2}$$

$$T_2 = T_1 \times \frac{V_2}{V_1} = 338.15 \times \frac{1.75}{2.00} = 295.9\text{K}$$

$$T_2 = 295.9 - 273.15 \approx 23\ °\text{C}$$

Hence, the room temperature is $23\ °\text{C}$.

3. A sample of nitrogen gas is placed in a mixture of ice and water at 1 atm pressure. The sample occupies 0.80 dm^3 of space. The same sample is transferred into boiling acetone where it occupies the volume of 0.96 dm^3. Calculate the boiling point of acetone?

The mixture of ice and water exist only at $0\,^{\circ}C$ i.e. 273.15K.

From Charles's law,

$$\frac{V_1}{T_1} = \frac{V_2}{T_2}$$

$$T_2 = T_1 \times \frac{V_2}{V_1} = 273.15 \times \frac{0.96}{0.80} \approx 328\text{K}$$

$$T_2 = 328 - 273.15 \approx 55\ °\text{C}$$

Thus, the boiling point of acetone is $55\ °\text{C}$.

4. A balloon tied on a ship moving from the Arctic Ocean to the Atlantic Ocean arrives Miami, USA. The temperature in the Arctic Ocean is $-2\,^{\circ}C$ and the corresponding volume is 1.50L. At Miami, the temperature is $28\,^{\circ}C$. Determine the volume occupied by the balloon at Miami?

From Charles's law,

$$\frac{V_1}{T_1} = \frac{V_2}{T_2}$$

$$V_2 = V_1 \times \frac{T_2}{T_1} = 1.50 \times \frac{28 + 273.15}{-2 + 273.15} = 1.5 = \times \frac{301.15}{271.15}$$

$$V_2 \approx 1.67L$$

Therefore, the volume occupied by the balloon at Miami is 1.67L.

Limitations of Charles's Law

1. The ideal gas equation is usually derived from the kinetic theory of gases, which presumes that molecules occupy negligible volume, do not attract each other and undergo elastic collisions (no loss of kinetic energy); an imaginary gas with exactly these properties is termed an ideal gas. The behavior of a real gas is close to that of an ideal gas under most circumstances, which makes the ideal gas law useful.
2. This law of volumes implies theoretically that as a temperature reaches absolute zero the gas will shrink down to zero volume. This is not physically correct, since in fact all gases turn into liquids at a low enough temperature, and Charles's law is not applicable at low temperatures for this reason.
3. The fact that the gas will occupy a non-zero volume - even as the temperature approaches absolute zero - arises fundamentally from the uncertainty principle of quantum theory. However, as the temperature is reduced, gases turn into liquids long before the limits of the uncertainty principle come into play due to the attractive forces between molecules which are neglected by Charles's Law.

Gay-Lussac's Law

Gay-Lussac's law is also known as pressure law or Amontons's law. The law correlates how the pressure of a gas increases with an increase in temperature. This law is named after French chemist Joseph Louis Gay-Lussac. He formulated this relationship in 1808. Gay-Lussac's law is similar to Charles's law, the only difference is the "volume" term in Charles's law is interchanged by the "pressure" term in Gay-Lussac's law.

For a fixed amount of an enclosed ideal gas, its pressure is directly proportional to its absolute temperature at a constant volume.

Formula: *P= constant x T.*

As the law states: The pressure and temperature of an ideal gas are directly proportional to each other at a constant volume and for a given mass of a gas. The statement can mathematically be expressed as:

$$P \propto T$$

By removing the proportionality, we get,

$$P = k \times T$$

where k is a constant of proportionality.

The above expression can be rearranged as:

$$\frac{P}{T} = k$$

The above expression is valid for a given mass of a gas and at a constant volume. Hence as the temperature increases, the pressure of the gas also increases and as the temperature decreases, the pressure decreases. It is clear from the above equation that the ratio of the pressure to temperature is independent of the pressure or temperature at a constant volume.

Note: In the above expression temperature should be in kelvin.

For a given amount ideal gas, P_1, T_1, and P_2, T_2, are the pressures and temperatures of condition 1 and condition 2 at a constant volume. From Gay-Lussac's law,

$$\frac{P_1}{T_1} = \frac{P_2}{T_2}$$

If the pressure of a gas obeying Gay-Lussac's law is double, then the temperature also gets double which is explained below.

$$P_2 = 2P_1$$

$$\frac{P_1}{T_1} = \frac{P_2}{T_2} = \frac{2P_1}{T_2}$$

$$T_2 = 2T_1$$

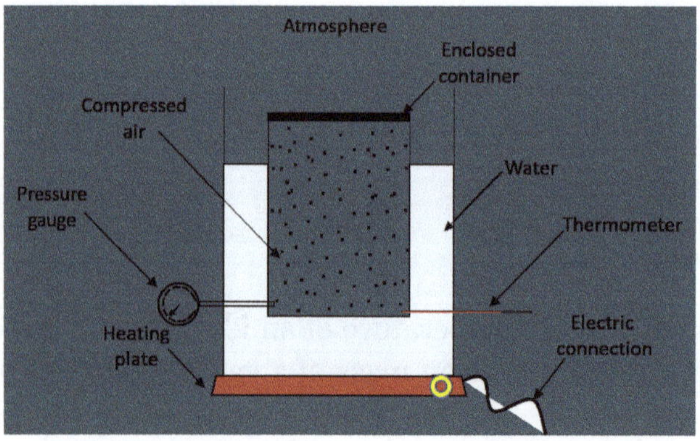

Figure: Experimental apparatus of Gay-Lussac's law.

The above diagram is a typical experimental setup required for verification of Gay-Lussac's law. When the air inside the chamber is heated with a heating source (electric heater is used in the diagram) via water, the temperature of the air increases i.e. the kinetic energy associated with the molecules of the air increases. Thus, molecules of the air exert more force in the outwards direction on all the walls of the container. Since the boundaries are rigid, the entrapped air inside the chamber is unable expands. The number of molecular collisions increases and the pressure of the air also increases. By the same analogy, when the air is cooled, the pressure of the air decreases.

In Charles's law, at least one of the boundaries of the vessel is movable. Thus, with the rise of temperature, the air inside the vessel expands. This expansion is what makes the pressure constant in Charles' law. While in Gay-Lussac's law the volume remains constant and the pressure varies.

Graphical Representation

The graphical representation of Gay-Lussac's law is demonstrated in the graphs.

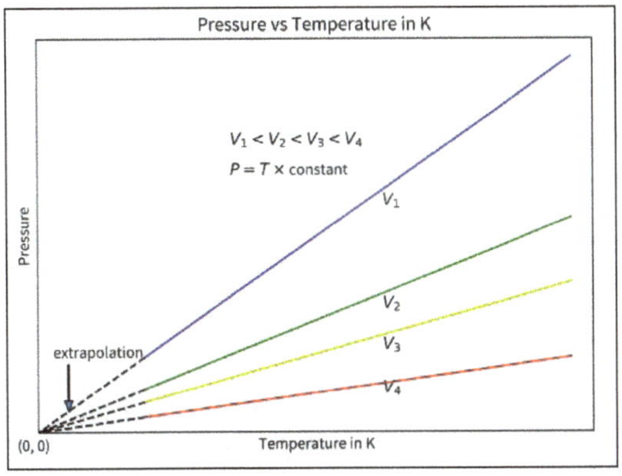

Figure: Pressure is linearly proportional to temperature. The graph is a straight line with positive slope and passes through the origin.

From the graph, the pressure increases linearly with an increase in the temperature. In the above figure, the temperature is taken in absolute scale. It can also be observed that as the temperature approaches zero, the pressure also approaches zero. All the above four lines are plotted at a constant volume (isochore).

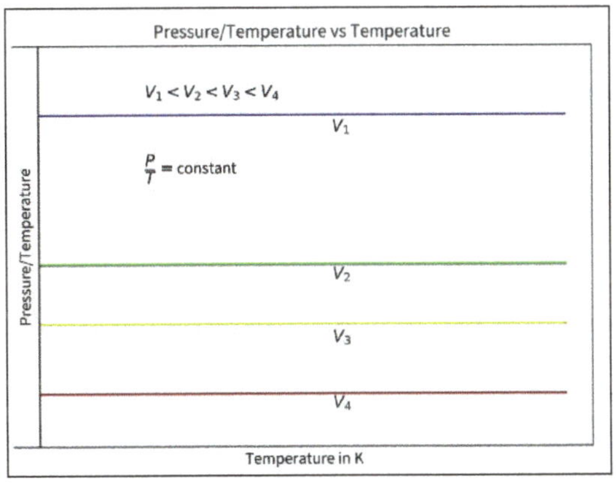

Figure: P/T is constant at a constant volume for a given amount of gas. The graph is a straight line parallel to x-axis.

The ratio of the pressure to temperature is independent of temperature (or pressure) at a constant volume. This can be observed from the above graph.

Limitations of Gay-Lussac's law

- The law is only applicable to ideal gases.
- Gay-Lussac's law holds good for real gases at high temperatures and/or low pressure.
- The ratio of the pressure to temperature deviated at high pressures. The ratio decreases with increasing the pressure. This decrease is due to an increase in volume at high pressures which is explained by an increase in repulsive forces among the molecules at high pressures.

Real World Applications of Gay-Lussac's Law

There are numerous applications of Gay-Lussac's law can be observed in day to day life. Below are some of the mentioned:

1. Pressure cooker: When a pressure cooker is kept on a heating source (stove). As per Gay-Lussac's law, the pressure of the fluid in the cooker increases with the rising of the temperature.
2. Aerosol can: An aerosol can on exposure to the sunlight or high temperatures, might bursts. The reason for bursting is increased in the pressure inside the can beyond the threshold limit. The pressure increases on the account of the rise of the temperature inside the can.
3. Gun bullet: When the bullet from a gun is ignited, the chemical energy stored in the shell of the bullet is converted into heat by chemical reactions. This heat increases the temperature which as per Gay-Lussac's law increases the pressure. Because of the high pressure, the bullet is fired from the gun.
4. Automobile tyres: The rupture of automobile tyres on subjection to high temperature is a classic example of Gay-Lussac's law. The high temperature pressurizes the air inside the tyres and beyond a point, they explode.

Gay-Lussac's law along with Boyles' law, Charles's law and Avogadro's law form ideal gas law.

Examples of Gay-Lussac's Law

1. Consider a fixed amount of butane gas at temperature $65\ ^{\circ}C$ at pressure $1.2 \times 10^4\ \mathrm{N\ m^{-2}}$ undergoes a change from stage 1 to stage 2 such that the initial and final volumes remain the same. The new temperature of the gas is $200\ ^{\circ}\mathrm{C}$. Calculate the new pressure?

First, convert both the temperatures in the problem statement from celsius to kelvin.

$$T_1 = 65 + 273.15 = 338.15\mathrm{K}$$

$$T_2 = 200 + 273.15 = 473.15\mathrm{K}$$

As from Gay-Lussac's law at constant volume and for a given mass of gas,

$$\frac{P_1}{T_1} = \frac{P_2}{T_2}$$

$$P_2 = \frac{P_1}{T_1} \times T_2 = \frac{1.4 \times 10^4}{338.15} \times 473.15 = 1.9 \times 10^4\ \mathrm{Nm^{-2}}$$

Therefore, the new pressure is $1.9 \times 10^4\ \mathrm{Nm^{-2}}$.

2. The pressure of water vapour in a pressure cooker is 112 700 N $\mathrm{m^{-2}}$. Estimate the boiling point of water in the pressure cooker. Assume the vapour behave ideally and obeys Gay-Lussac's law.

We all know that the boiling point of water at 1 atmospheric pressure is $100\ ^{\circ}C$. And 1 atmosphere is equivalent to 112 700 N m^{-2}. Thus, applying Gay-Lussac's law,

$$\frac{P_1}{T_1}=\frac{P_2}{T_2}$$

$$T_2=\frac{P_2}{P_1}\times T_1=\frac{112\,700}{101\,325}\times(100+273.15)$$

$$T_2=\frac{112\,700}{101\,325}\times 373.15=415.04\text{K}\approx 142\ ^{\circ}\text{C}$$

The boiling point of the water inside the pressure cooker is $142\ ^{\circ}\text{C}$.

3. A soda bottle at the room the temperature of $25\ ^{\circ}C$ and the pressure of 2atm is heated to the temperature of $330\ ^{\circ}\text{C}$ at which it bursts. Calculate the pressure of the heated soda bottle. Converting the initial and final temperatures are from degree celsius to kelvin.

$$T_1=25+273.15=298.15\text{K}$$

$$T_1=330+273.15=603.15\text{K}$$

From the Gay-Lussac's law,

$$\frac{P_1}{T_1}=\frac{P_2}{T_2}$$

$$P_2=\frac{P_1}{T_1}\times T_2=\frac{2}{298.15}\times 603.15\approx 4.05\ \text{atm}$$

The soda bottle bursts at the pressure of 4.05 atm.

Avogadro's Law

Avogadro's law is also known as Avogadro's hypothesis or Avogadro's principle. The law dictates the relationship between the volume of a gas to the number of molecules the gas possesses. This law like Boyle's law, Charles's law, and Gay-Lussac's law is a specific case of the ideal gas law. This law is named after Italian scientist Amedeo Avogadro. He formulated this relationship in 1811. After conducting the experiments, Avogadro hypothesized that the equal volumes of gas contain the equal number of particles. The hypothesis also reconciled Dalton atomic theory. In 1814 French Physicist Andre-Marie Ampere published similar results. Hence, the law is also known as Avogadro-Ampere hypothesis.

For an ideal gas, equal volumes of the gas contain the equal number of molecules (or moles) at a constant temperature and pressure.

In other words, for an ideal gas, the volume is directly proportional to its amount (moles) at a constant temperature and pressure.

Formula: $V = constant \times n$

As the law states: volume and the amount of gas (moles) are directly proportional to each other at constant volume and pressure. The statement can mathematically express as:

$$V \propto n$$

Replacing the proportionality,

$$V = k \times n$$

where k is a constant of proportionality.

The above expression can be rearranged as:

$$\frac{V}{n} = k$$

The above expression is valid for constant pressure and temperature. From Avogadro's law, with an increase in the volume of a gas, the number of moles of the gas also increases and as the volume decreases, the number of moles also decreases.

If V_1, V_2 and n_1, n_2 are the volumes and moles of a gas at condition 1 and condition 2 at constant temperature and pressure, then using Avogadro's law we can formulate the below equation.

$$\frac{V_1}{n_1} = \frac{V_2}{n_2}$$

Let the volume V_2 at condition 2 be twice the volume V_1 at condition 1.

$$V_2 \quad V_1$$

$$\frac{V_1}{n_1} = \frac{2V_1}{n_2}$$

$$n_2 = 2n_1$$

Therefore, with doubling the volume, the number of moles also gets double.

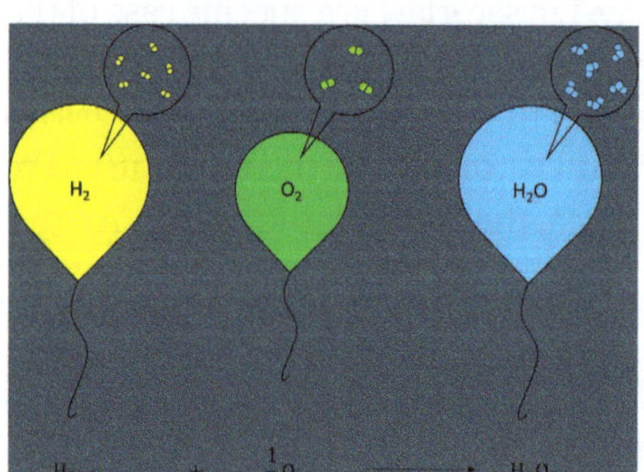

Figure: Avogadro's law (Combination of hydrogen and oxygen to form water vapour).

In the diagram, 1 mol (n_{H2}) of hydrogen gas reacts with a $\frac{1}{2}$ mol(n_{O_2}) of oxygen gas to form 1 mol(n_{H_2O}) of water vapour. The consumption of hydrogen is twice the consumption of oxygen which is expressed below as:

$$n_{H_2} = 2n_{O_2} = n_{H_2O}$$

Let say, 1 mol of hydrogen occupies volume V_{H_2}, $\frac{1}{2}$ mol of oxygen occupies V_{O_2} and similarly for 1 mol of water vapour, volume V_{H_2O}. As we know from Avogadro's law, equal volumes contain equal moles. Hence, the relationship between the volumes is the same as among the moles as follows:

$$V_{H_2} = 2V_{O_2} = V_{H_2O}$$

Avogadro's law along with Boyles' law, Charles's law and Gay-Lussac's forms ideal gas law.

Graphical Representation

The graphical representation of Avogadro's law is shown below.

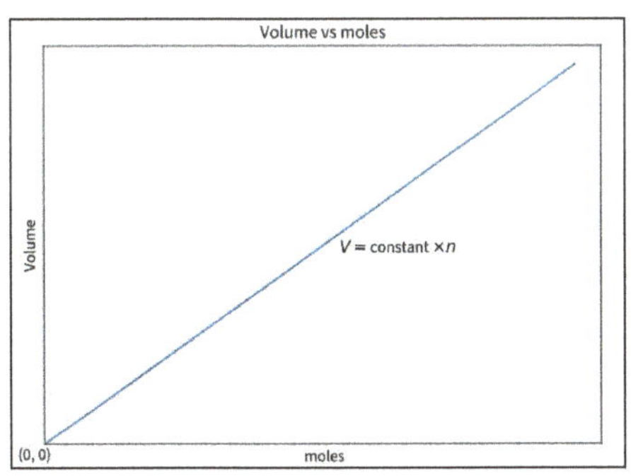

Figure: Volume vs Mole Graph

The above graph is plotted at constant temperature and pressure. As we can observe from the graph that the volume and mole have a linear relationship with the line of a positive slope passing through the origin.

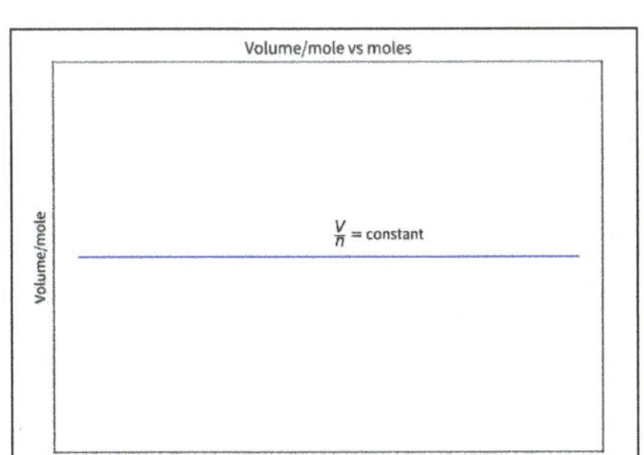

Figure: Volume/Mole vs Mole Graph

As shown in the above figure, line $\frac{V}{n} = k$ is parallel to the x-axis. It means that the value of volume by mole is constant and is not influenced by any change in mole (or volume).

Both the above graphs are plotted at a constant temperature and pressure.

Avogadro's Constant

The Avogadro's constant is a constant named after Avogadro, but Avogadro did not discover it. The Avogadro's constant is a very useful number; the number defines the number of particles constitutes in any material. It is denoted by N_A and has dimension mol^{-1}. Its approximate value is given below.

$$N_A \approx 6.022 \times 10^{23} \text{mol}^{-1}$$

Molar Volume

Since Avogadro's law deals with the volume and moles of a gas, it is necessary to discuss the concept of molar volume. The molar volume as from the name itself is defined as volume per mole. It is denoted as V_m and having a unit of volume divided by a unit of mole (e.g. $dm^3 mol^{-1}, m^3 kmol^{-1}, cm^3 mol^{-1}$ etc). From the ideal gas law, at STP ($T = 273.15K, P = 101325Pa$) the molar volume is calculated as:

$$V_m = \frac{V}{n} = \frac{RT}{P} = \frac{8.314 \times 273.15}{101325}$$

$$= 0.02241 m^3 mol^{-1}$$

$$= 22.42 dm^3 mol^{-1}$$

Limitation of Avogadro's Law

- The law works perfectly only for ideal gases.
- The law is approximate for real gases at low pressure and/or high temperature.
- At low temperature and/or high pressure, the ratio of volume to mole is slightly more for real gases compare to ideal gases. This is because of the expansion of real gases due to intermolecular repulsion forces at high pressure.
- Lighter gas molecules like hydrogen, helium etc., obey Avogadro's law better in comparison to heavy molecules.

Real World Applications of Avogadro's Law

Avogadro's principle is easily observed in everyday life. Below are some of the mentioned:

- Balloons: When you blow up a balloon, you are literally forcing the air from your mouth to inside the balloon. In other words, you are filling more moles of air in the balloon and it expands.
- Tyres: Have you ever filled deflated tyres? If yes, then you are nothing but following Avogadro's law. When you pump air inside the deflated tyres at a gas station, the amount (moles) of gas inside the tyres is increased which increases the volume and the tyres are inflated.

- Human lungs: When we inhale, air flows inside our lungs and they expand while when we exhale, the air flow from the lungs to surroundings and the lungs shrink.

Laboratory Experiment to Prove Avogadro's Law

Objective

To verify Avogadro's law by estimating the amount (moles) of different gases at a fixed volume, temperature and pressure.

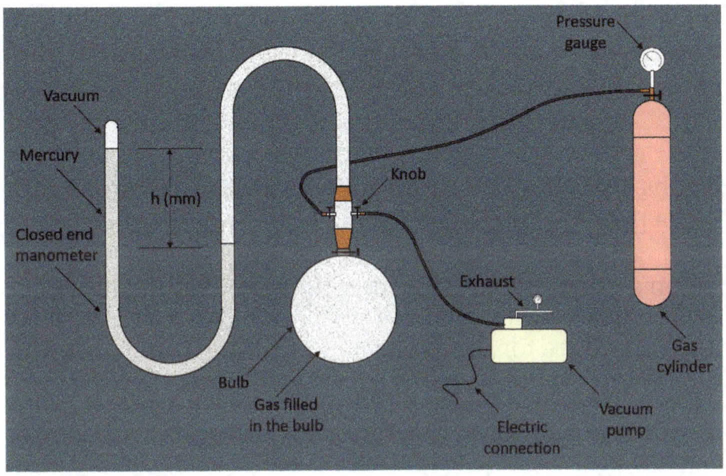

Figure: Experimental apparatus for verification of Avogadro's law.

Apparatus

The apparatus requires for this experiment is shown in the above diagram. It consists of a U-tube manometer (in the diagram closed-end manometer is used, but opened-end manometer can also be used) as depicted in the figure, mercury, a bulb, a vacuum pump, four to five cylinders of different gases and a thermometer. Connect the all apparatuses as shown in figure.

Nomenclature

1. V_0 is the volume of the bulb, which is known (or determined) before the experiment.
2. T is the temperature at which the experiment is performed, which can be determined from the thermometer (for simplicity take it as room temperature).
3. P is the pressure at which the experiment is performed, which can be determined from the difference in heights of mercury level in the manometer.
4. W_0 is the empty weight of the bulb, and it is known (or determined) before the experiment.
5. W is the filled weight of the bulb.
6. W_g is the weight of the gas inside the bulb.
7. M is the molecular weight of the gas.

Procedures

1. Take a gas cylinder attached it the bulb setup and also attached the pump to the bulb set-up. Care must be taken while attaching the apparatus to prevent any leakages of the gas.
2. First, close the knob of the gas cylinder and open the vacuum pump knob on the bulb. Evacuate the air filled in the system and by turning on the vacuum pump.
3. Once the bulb is emptied, close the vacuum pump knob and switch off the vacuum pump.
4. Start filling the bulb with the cylinder gas by opening the gas cylinder knob slowly until the desired difference in the mercury height is achieved. Note the height difference in the manometer. (The value of the height difference should be the same for all the readings.)
5. Close all the knobs also close the connection between the bulb and the manometer to isolate the gas inside the bulb. Disassemble the bulb from the manometer.
6. Weigh the bulb on a weighing machine and note the reading down.
7. This finishes the procedure for the first gas. Repeat the same procedure for different gases.

Calculation

Calculate the weight of gas (W_g) in the bulb by subtracting the weight of empty bulb (W_0) from the weight of the filled bulb (W).

$$W_{\mathrm{g}} = W_0 - W$$

Then calculate the number of moles of the gas as:

$$n_{\mathrm{g}} = \frac{W_{\mathrm{g}}}{M}$$

The number of moles of all gases should be approximately equal within a small percentage of error. If this is true, then all the gases do obey the Avogadro's law.

If the experiment is performed at STP ($T = 273.15K, P = 101325Pa$), then we can also calculate the molar volume V_m as:

$$V_{\mathrm{m}} = \frac{V_0}{n_{\mathrm{g}}}$$

And its value should be close to $22.4\mathrm{dm}^3\mathrm{mol}^{-1}$.

Examples to Prove Avogadro's Law

1. Consider 20 mol of hydrogen gas at temperature 0 °C and pressure 1 atm having the volume of $44.8\ \mathrm{dm}^3$. Calculate the volume of 50 mol of nitrogen gas, at the same temperature and pressure?

As from Avogadro's law at constant temperature and pressure,

$$\frac{V_1}{n_1} = \frac{V_2}{n_2}$$

$$V_2 = \frac{V_1}{n_1} \times n_2 = \frac{44.8}{20} \times 50 = 112\ \text{dm}^3$$

Therefore, the volume is 112 dm^3.

2. There is the addition of 2.5 L of helium gas in 5.0 L of helium balloon; the balloon expands such that pressure and temperature remain constant. Estimate the final moles of gas if the gas initially possesses 8.0 mol.

The final volume is the addition of the initial volume and the volume added.

$$V_2 = 2.5 + 5.0 = 7.5\ \text{L}$$

From Avogadro's law,

$$\frac{V_1}{n_1} = \frac{V_2}{n_2}$$

$$n_2 = n_1 \times \frac{V_2}{V_1} = 8.0 \times \frac{7.5}{5.0} = 12\ \text{mol}$$

The final number of moles in 7.5 L of the gas is 12 mol.

3. 3.0 L of hydrogen reacts with oxygen to produce water vapour. Calculate the volume of oxygen consumed during the reaction (assume Avogadro's law holds)?

For the consumption of every one mole of hydrogen gas, half a mole of oxygen is consumed.

$$n_{O_2} = \frac{1}{2} n_{H_2}$$

As per Avogadro's law, the volume is directly proportional to moles, so we can rewrite the above equation as:

$$V_{O_2} = \frac{1}{2} V_{H_2} = \frac{1}{2} \times 3.0 = 1.5\ \text{L}$$

1.5 L of oxygen is consumed during the reaction.

Ideal Gas Equations

The relations known as Boyle's law, Charles' law, and Avogadro's law can be combined into an exceedingly useful formula called the Ideal Gas Equation,

$$PV = nRT$$

where R denotes the gas constant:

$$R = 0.082 \frac{\text{liter - atm}}{\text{K - mole}}$$

The temperature is, as always in gas equations, measured in Kelvin.

This formula is strictly valid only for ideal gases—those in which the molecules are far enough apart so that intermolecular forces can be neglected. At high pressures, such forces cause significant departure from the Ideal Gas Equation, and more complicated equations have been devised to treat such cases. The Ideal Gas Equation, however, gives useful results for most gases at pressures less than 100 atmospheres.

$$V = \frac{nRT}{P}$$

$$V = \frac{(1\,mole)\left(\dfrac{0.082\,\text{L - atm}}{\text{K - mole}}\right)(273.15\,K)}{1\text{ atm}}$$

$$V = 22.40\text{ liters}$$

This is the value stated in the carbon dioxide reaction; you were asked to memorize that 1 mole of any gas occupies 22.4 liters at STP.

You should be able to use the Ideal Gas Equation to determine any one of the four quantities—pressure, volume, moles, or temperature—if you are given values for the other three.

One important application is to deduce the molecular mass and formula for a gas. Assume that you know that the hydrocarbon propylene is, by mass, 85.6% carbon and 14.4% hydrogen. Then the atomic ratios of the compound are:

$$C = \frac{85.6}{12.01} = \frac{7.13}{1}$$

$$H = \frac{14.4}{1.01}$$

$$\frac{H}{C} = \frac{14.26}{7.13} = \frac{2}{1}$$

Therefore, the propylene molecule is some integral multiple of CH_2: It can be CH_2 or C_2H_4 or C_3H_6 or a yet larger molecule. Measuring the volume of 10 grams of propylene at STP yields 5.322 liters, which you can use to calculate its molecular mass.

$$\frac{5.322\text{ liters}}{22.40\text{ liters/ mole}} = 0.2376\text{ mole}$$

$$\frac{10\text{ grams}}{0.2376\text{ mole}} = 42.09\text{ grams/ mole (the molecular mass)}$$

Because the atomic masses of 1 CH_2 unit added together is 14.03, the molecule contains three such units. Consequently, the molecular formula for propylene is C_3H_6.

- What is the volume occupied by 1 kilogram of carbon monoxide at 700 °C and 0.1 atm?
- The ozone molecule contains only oxygen atoms. Determine the molecular formula of ozone given that 2.3 grams occupies 1,073 milliliters at standard temperature and pressure.

Ideal Gas Equation Units

The Ideal gas equation units, terms and corresponding definitions can be expressed as follows-

Terms	Symbol	Definition	Units
Pressure	P	Force per units area.	Pa or N/m^2
Volume	V	Three-dimensional space enclosed by the closed surface.	m^3
Amount of substances/ number of moles	n	The ratio of the mass of the gas(m) to its molar mass(M).	mole
Ideal gas constant	R	Physical constant relating the average kinetic energy of an ideal gas with temperature. It is expressed in units of energy per temperature increment per mole.	8.3144598(48) $J.K^{-1}.mol^{-1}$
Temperature	T	Measure of heat.	K or OC

Real Gas

Real gases can deviate from ideal behaviour, especially at *high pressures* and *low temperatures*. The extent of deviation is measured using the compressibility factor. The compressibility factor is obtained by solving for *n* in the ideal gas law: dividing the product of pressure and volume by the product of the gas constant and temperature (PV/RT) for one mole of a given substance. Under ideal conditions, this ratio of PV/RT should be exactly equal to 1.

Real Gases at High Pressure

At higher pressures, gas molecules are closer together in a space. As a result of this crowding, gas molecules experience greater attractive intermolecular forces. Intermolecular forces hold molecules together more, lessening the force and frequency of collisions with the container wall and thus lowering the pressure below ideal values. As well, at higher pressure, molecules occupy a larger proportion of the volume of the container. With other gas molecules taking up a larger proportion of the volume of the container, the unoccupied volume of the container available to any one molecule is smaller than in ideal conditions. This decrease in available volume causes an increase in pressure beyond ideal conditions.

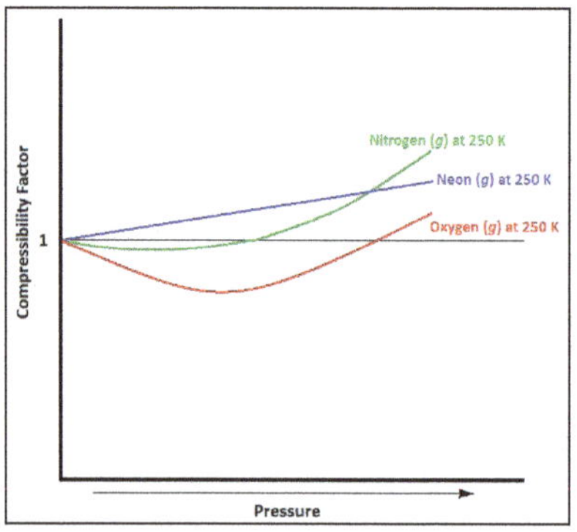

Figure: Approximate compressibility factors of three gases at 250 K.

Real Gases at Low Temperature

Temperature also influences deviations from ideal gas behaviour. As temperature decreases, the average kinetic energy of the gas particles decreases. A larger proportion of gas molecules therefore have insufficient kinetic energy to overcome attractive intermolecular forces from neighbouring atoms. This means that gas molecules become "stickier" to each other, and collide with the walls of the container with less frequency and force, decreasing pressure below that of ideal values.

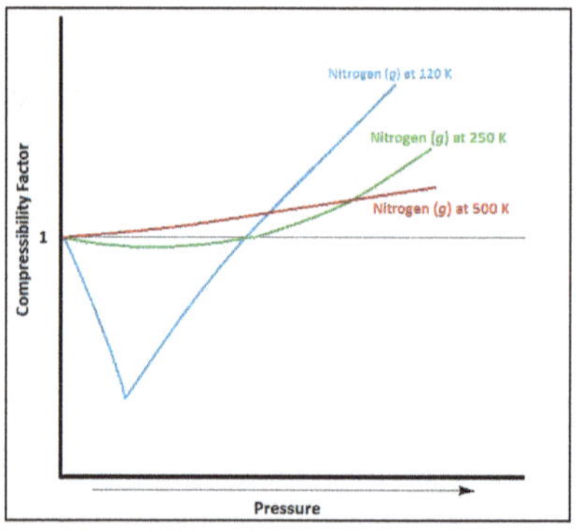

Figure: Approximate compressibility factor of nitrogen at different temperatures.

Real Gas Equations

To understand the behaviour of real gases, the following must be taken into account:

- Compressibility effects;
- Variable specific heat capacity;
- Van der Waals forces;
- Non-equilibrium thermodynamic effects;
- Issues with molecular dissociation and elementary reactions with variable composition.

For most applications, such a detailed analysis is unnecessary, and the ideal gas approximation can be used with reasonable accuracy. On the other hand, real-gas models have to be used near the condensation point of gases, near critical points, at very high pressures, to explain the Joule–Thomson effect and in other less usual cases.

Models

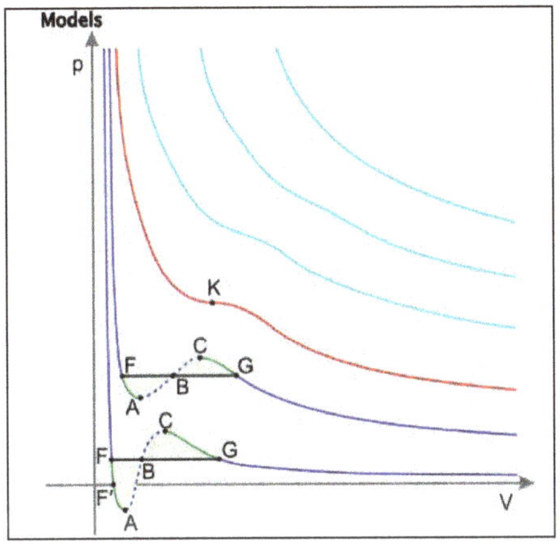

Isotherms of Real Gas (Sketchy)

- Dark blue curves – Isotherms below the critical temperature. Green sections – metastable states.
- The section to the left of point F – Normal liquid.
- Point F – Boiling point.
- Line FG – Equilibrium of liquid and gaseous phases.
- Section FA – Superheated liquid.
- Section F′A – Stretched liquid ($p< 0$).
- Section CG – Supercooled vapor.
- Point G – Dew point.
- The plot to the right of point G – Normal gas.
- Areas FAB and GCB are equal.
- Red curve – Critical isotherm.
- Point K – Critical point.
- Light blue curves – Supercritical isotherms.

Van der Waals Model

Real gases are often modeled by taking into account their molar weight and molar volume:

$$RT = \left(P + \frac{a}{V_m^2}\right)(V_m - b)$$

Where P is the pressure, T is the temperature, R the ideal gas constant, and V_m the molar volume. a and b are parameters that are determined empirically for each gas, but are sometimes estimated from their critical temperature (T_c) and critical pressure (P_c) using these relations:

$$a = \frac{27R^2T_c^2}{64P_c}$$

$$b = \frac{RT_c}{8P_c}$$

Redlich–Kwong Model

The Redlich–Kwong equation is another two-parameter equation that is used to model real gases. It is almost always more accurate than the van der Waals equation, and often more accurate than some equations with more than two parameters. The equation is:

$$RT = P(V_m - b) + \frac{a}{V_m(V_m + b)T^{\frac{1}{2}}}(V_m - b)$$

where a and b two empirical parameters that are not the same parameters as in the van der Waals equation. These parameters can be determined:

$$a = 0.4275\frac{R^2T_c^{2.5}}{P_c}$$

$$b = 0.0867\frac{RT_c}{P_c}$$

Berthelot and Modified Berthelot Model

The Berthelot equation (named after D. Berthelot) is very rarely used,

$$P = \frac{RT}{V_m - b} - \frac{a}{TV_m^2}$$

but the modified version is somewhat more accurate:

$$P = \frac{RT}{V_m}\left[1 + \frac{9P/P_c}{128T/T_c}\left(1 - \frac{6}{(T/T_c)^2}\right)\right]$$

Dieterici Model

This model (named after C. Dieterici) fell out of usage in recent years,

$$P = RT\frac{\exp\left(\frac{-a}{V_mRT}\right)}{V_m - b}$$

Clausius Model

The Clausius equation (named after Rudolf Clausius) is a very simple three-parameter equation used to model gases,

$$RT = \left(P + \frac{a}{T(V_m + c)^2}\right)(V_m - b)$$

Where,

$$a = \frac{27R^2T_c^3}{64P}$$

$$b = V_c - \frac{RT_c}{4P_c}$$

$$c = \frac{3RT_c}{8P_c} - V_c$$

where V_c is critical volume.

Virial Model

The Virial equation derives from a perturbative treatment of statistical mechanics.

$$PV_m = RT\left(1 + \frac{B(T)}{V_m} + \frac{C(T)}{V_m^2} + \frac{D(T)}{V_m^3} + \ldots\right)$$

or alternatively:

$$PV_m = RT\left(1 + \frac{B'(T)}{P} + \frac{C'(T)}{P^2} + \frac{D'(T)}{P^3} + \ldots\right)$$

where A, B, C, A′, B′, and C′ are temperature dependent constants.

Peng–Robinson Model

Peng–Robinson equation of state (named after D.-Y. Peng and D. B. Robinson) has the interesting property being useful in modeling some liquids as well as real gases,

$$P = \frac{RT}{V_m - b} - \frac{a(T)}{V_m(V_m + b) + b(V_m - b)}$$

Wohl Model

The Wohl equation (named after A. Wohl) is formulated in terms of critical values, making it useful when real gas constants are not available,

$$RT = \left(P + \frac{a}{TV_m(V_m + b)} - \frac{c}{T^2V_m^3}\right)(V_m - b)$$

Where,

$$a = 6P_cT_cV_c^2$$

$$b = \frac{V_c}{4}$$

$$c = 4P_cT_c^2V_c^3$$

Beattie–Bridgman Model

This equation is based on five experimentally determined constants. It is expressed as,

$$P = \frac{RT}{v^2}\left(1 - \frac{c}{vT^3}\right)(v + B) - \frac{A}{v^2}$$

Where,

$$A = A_0\left(1 - \frac{a}{v}\right)$$

$$B = B_0\left(1 - \frac{b}{v}\right)$$

This equation is known to be reasonably accurate for densities up to about 0.8 ρcr, where ρcr is the density of the substance at its critical point. The constants appearing in the above equation are available in following table when P is in KPa, v is in $\frac{m^3}{K\ mol}$, T is in K and R=8.314 $\frac{kPa.m^3}{K\ mol.K}$.

Gas	Ao	a	Bo	b	c
Air	131.8441	0.01931	0.04611	-0.001101	4.34×10^4
Argon, Ar	130.7802	0.02328	0.03931	0.0	5.99×10^4
Carbon Dioxide, CO_2	507.2836	0.07132	0.10476	0.07235	6.60×10^5
Helium, He	2.1886	0.05984	0.01400	0.0	40
Hydrogen, H_2	20.0117	-0.00506	0.02096	-0.04359	504
Nitrogen, N_2	136.2315	0.02617	0.05046	-0.00691	4.20×10^4
Oxygen, O_2	151.0857	0.02562	0.04624	0.004208	4.80×10^4

Benedict–Webb–Rubin Model

The BWR equation, sometimes referred to as the BWRS equation,

$$P = RTd + d^2\left(RT(B + bd) - (A + ad - a\alpha d^4) - \frac{1}{T^2}[C - cd(1 + \gamma d^2)\exp(-\gamma d^2)]\right)$$

where d is the molar density and where a, b, c, A, B, C, α, and γ are empirical constants. Note that the γ constant is a derivative of constant α and therefore almost identical to 1.

References

- Definition-of-ideal-gas: thoughtco.com, Retrieved 29 August, 2019
- Ideal-Bose-gas: uni-frankfurt.de, Retrieved 9 February, 2019
- Ideal-Fermi-gas: uni-frankfurt.de, Retrieved 17 January, 2019
- Gas-laws, states-of-matter, chemistry: toppr.com, Retrieved 13 June, 2019
- Boyle-law: chemistrygod.com, Retrieved 4 March, 2019
- Charles-law: chemistrygod.com, Retrieved 15 February, 2019
- Gay-lussacs-law: chemistrygod.com, Retrieved 7 July, 2019
- Avogadros-law: chemistrygod.com, Retrieved 12 March, 2019
- Ideal-gas-equation, gases, chemistry: cliffsnotes.com, Retrieved 2 August, 2019
- Ideal-gas-equation, physics: byjus.com, Retrieved 22 April, 2019
- Real-gases: opentextbc.ca, Retrieved 10 January, 2019
- Quantum-Mechanics-Real-Gases, chemical-engineering: idc-online.com, Retrieved 21 May, 2019

3

Surface Chemistry

The branch of chemistry which deals with the phenomena which occur at the surfaces of substances is known as surface chemistry. Some of these phenomena are adsorption and molecular diffusion. The topics elaborated in this chapter will help in gaining a better perspective about the topics which are studied under surface chemistry.

The field of surface chemistry has undergone an explosive development in the period marked by its two Nobel prizes—1932 (Irving Langmuir) to 2007 (Gerhard Ertl). Indeed, many years went by after Langmuir before modern surface measurement techniques began to develop significantly (in the 1960s) as a result of the need to understand semiconductor surface chemistry and physics for applications to emerging electronic devices. The period after 1960 witnessed the development of ultrahigh vacuum techniques for probing surfaces using electrons, ions, photons, and heat as probes of chemistry on single-crystal metal, semiconductor, and insulator surfaces. These new measurement methods were often the product of the work of surface physicists, and the term surface science began to be used to encompass developments in the field that mixed chemistry and physics.

In parallel to the development of the fundamental basis of surface chemistry, there has been a tremendous impact of this knowledge on technology of many sorts. At the beginning of the era bounded by Langmuir and Ertl, the improvement of the incandescent lamp was a major technological driver, and the interest in the chemistry on tungsten filaments spread quickly to related interests in the behavior of molecules adsorbed and reacting on transition metal surfaces, forming the underpinnings of fundamental heterogeneous catalysis. As interest in semiconductor surfaces blossomed in the 1960s, fundamental understanding of the role of surface structure affected both catalyst and semiconductor science. For many years, a substantial portion of the world's GDP has involved surface chemistry, and the wide range of high-impact applications of surface chemistry has been a driver in the development of the field. Among the most important are the electronics industry, the petrochemical industry, the conversion of air into fixed-nitrogen substances such as fertilizers, and the automotive catalyst technology. Emerging areas include the photovoltaic industry and the wide range of energy conversion technologies.

Modern surface chemistry is about molecular-level understanding and the control of surface chemical reactions. Over the past decades, various surface science techniques have been developed and a vast amount of knowledge about surface chemistry has been accumulated. The fundamental knowledge that is accumulated provides a foundation for the development of many industrial technologies that produce items including chemicals and fuels, semiconductor devices and biomedical devices. Technological advances, in turn, necessitate the further development of new surface characterization techniques with higher spatial, temporal, and energy resolution.

This perspective begins with an overview of several applications of surface chemistry that have an enormous economic impact on our society. Advances in the fundamental understanding of molecular processes on surfaces over the past decades are then reviewed with an emphasis on the development of the molecular-level surface techniques. We highlight new scientific insights provided by in situ experimental and theoretical techniques capable of characterizing surfaces and interfaces under working conditions.

Applications of Surface Chemistry

Surface chemistry holds a very special position in the development and operations of many technologies that will be critical for our energy and environmental future. Thin films must be designed and grown for everything from solar cells, microelectronics, computer chips, chemical and biochemical sensors, prosthetic medical devices, reflective and protective coatings, optical, electro-optic, and opto-electric devices, and adhesives. Surface chemistry is also central to the preparation and operation of sorbents, solid reactants, catalysts for clean fuels and chemicals production, pollution cleanup, photocatalysts, fuel cells, and batteries. The synthesis, stabilization, and use of all forms of nanomaterial also involve a great deal of surface chemistry. With the advances made within the past decade in fundamental understanding of surface chemistry, countless as yet unknown applications can also be imagined.

Throughout modern industrialization, surface chemistry plays an indispensable role in various industrial technologies for chemical and energy conversion, information processing, heath care, and material and environmental protection. The paramount importance of surface chemistry is reflected in the tremendous economic impact made by these technologies. The roles of surface chemistry in several major industrial technologies are discussed here. The major challenges for surface chemistry in further technological developments are highlighted.

Heterogeneous Catalyst Technology

A catalyst is an entity that accelerates a chemical reaction without being consumed in the process. This ability is usually referred to as the activity of a catalyst. For a chemical reaction with multiple possible products, a catalyst may promote the production of a specific product, which is referred to as the selectivity of a catalyst. Heterogeneous catalytic reactions occur on the surface of solid catalysts and involve elementary surface chemical processes such as adsorption of reactants from a reaction mixture, surface diffusion and reaction of adsorbed species, and desorption of reaction products. The acceleration of a chemical reaction is due to the high reactivity of surface atoms that facilitates bond breaking and bond rearrangement of adsorbed molecules.

In the late 1800s, the large-scale synthesis of simple but important bulk chemicals, including sulfuric acid, ammonia, and nitric acid, were among the first industrial processes based on heterogeneous catalysis. After these early successes, many complex catalytic processes evolved in the chemical industry. These processes include the synthesis of methanol from carbon monoxide and hydrogen, the oxidation of ethylene to ethylene oxide for the manufacture of ethylene glycol antifreeze, the dehydrogenation of butane to butadiene for the production of synthetic rubber, and the polymerization of ethylene and propylene to produce plastics.

In the middle of the 1900s, the large-scale industrial applications of heterogeneous catalysis emerged in the petroleum industry with the development of catalytic cracking processes for naphtha refining. Since then, the applications of zeolites as cracking catalysts, the supported metal clusters as reforming catalysts, in the petroleum industry have allowed large-scale production of high quality fuels and bulk chemicals. These developments have helped to lay out the foundation of our current fast-growing, petroleum-based economy.

The development of catalytic processes for pollution control and prevention also accompanied the fast-growing economy. One of many successful examples of pollution control is the three-way catalytic converter that reduces the pollutant emission from automobiles shows that, despite a twofold increase in the number of automobiles, mobile pollutant emissions in the Los Angeles area have continuously decreased since the three-way catalytic converter was introduced in 1975.

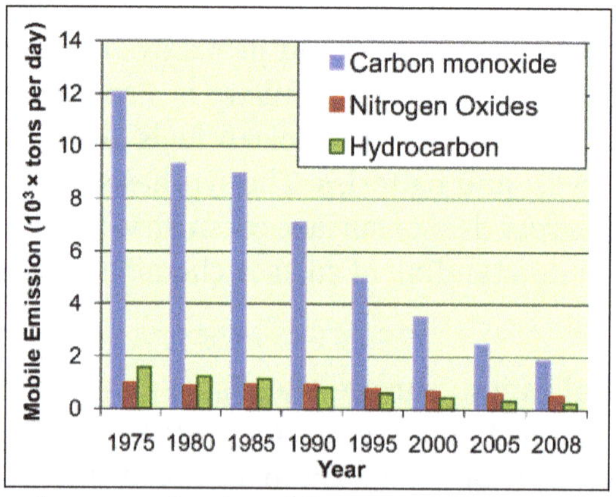

Figure: Annual average of toxic mobile emissions in Los Angeles County.

In the 21st century, emphasis shifted to catalytic selectivity. It is desired to produce one specific product out of many other thermodynamically possible compounds that are undesirable waste byproducts (green chemistry). To identify the molecular factors that control selectivity, it is required to monitor surface chemical processes under reaction conditions. Innovations in catalytic processes are playing a key role in green chemistry by reducing or eliminating the use and generation of hazardous substances in chemical conversion processes and in the development of renewable energy and chemical feedstocks. An example is the commercialization of the hydrogen peroxide to propylene oxide process for making propylene oxide, a polyurethane raw material. In this process, propylene is oxidized by hydrogen peroxide over a Ti-silicalite catalyst. The use of toxic reagents, such as chlorine, and the production of coproducts, such as chlor–alkali complexes, *tert*-butyl alcohol, or styrene, are avoided.

Many challenging catalytic processes are under intensive research and development for:

1. Efficient water and co_2 splitting to produce fuels by photocatalysts,
2. Efficient conversion of fuels to electricity by electrochemical fuel cells,
3. Selective conversion of biomass to fuels and bulk chemicals,
4. Selective production of fine chemicals and pharmaceutical compounds by heterogeneous catalysts and
5. Greenhouse gas (e.g., CO_2 and CH_4) reduction and utilization in fuel and chemical synthesis.

To achieve the optimal catalytic properties (efficient sunlight and electrical energy conversion, and high activity and selectivity for chemical conversion), much of this research is heavily relying on controlling the structure and composition of catalyst systems on a nanometer or molecular scale (nanocatalysis). For example, the catalytic activity and selectivity of a transition metal nanocatalyst can be tuned by controlling the size and shape of the nanocatalyst.

Semiconductor-based Technology

The transistor and the integrated circuit shaped the semiconductor industry and the modern information-driven world. A transistor is a semiconductor device used to amplify or switch an electronic signal, and an integrated circuit contains millions of transistors etched out of a small semiconductor surface. Integrated circuits are in many modern electronic devices such as computers and cell phones.

Extensive research on the effect of electron surface states led to the creation of the first point-contact germanium transistor in 1947. After the invention of transistors, advances in epitaxial growth of silicon thin films, chemical vapor deposition, and surface etching techniques made the industrial large-scale manufacture of silicon-based integrated circuits possible. Since then, the number of transistor that can be placed inexpensively on an integrated circuit has doubled approximately every 2 y, improving the performance and lowering the manufacturing cost of integrated circuits. New 18-nm manufacturing processes are expected to arrive in 2018. These processes include the following: the growth and deposition of sub–10-nm thin films of low-k insulating materials (for example, SiOC) and high-k gate dielectric materials (for example, SiO_xN_y); the conformal etching of deposited thin films with high aspect ratios; high-spatial-resolution patterning with new lithography techniques; the doping of the sub–10-nm silicon channels; and the deposition of new wiring materials with low resistance and high reliability (for example, Cu). The development of these new manufacturing processes leads to a broad range of surface and interface chemistry issues.

Unlike the silicon semiconductor, compound semiconductors such as GaAs, GaN, and InGaN are optically active materials. Their tunable optical response through composition control makes them extremely useful for optoelectronics applications. Other advantages of compound semiconductors over silicon include the following: higher electron mobility and electron velocity, low noise, and higher breakdown voltage, which make them ideal for radio frequency communication devices such as cell phones and wireless local area network cards. However, their compound character often introduces formidable surface chemistry challenges in large-scale manufacturing processes. For example, after nearly 30 y of intensive research, the discovery of a metal–organic chemical vapor deposition process (MOCVD) for growing device-quality GaN thin films finally enabled the commercialization of the GaN-based technology. In this heteroepitaxial growth process, a thin AlN buffer layer is deposited on the substrate before the GaN growth. The essential role of this buffer layer is to serve as a template for nucleation and to promote lateral growth of the GaN film due to the decrease in interfacial free energy between the film and the substrate.

Medical Technology

In the late 1940s, along with the development of biomaterials for medical implants such as intraocular lenses, hip joint replacement, and blood contact devices, researchers began exploring biological reactions at surfaces. It was indicated by many observations that the biocompatibility of a material is correlated with the surface properties of the material. For example, when a foreign material comes into contact with blood, it will rapidly adsorb various proteins onto its surface.

This process is usually referred to as nonspecific protein adsorption. The adsorbed protein layer determines subsequent platelet adhesion, which plays a major role in thrombogenesis on foreign surfaces. To prevent thrombogenesis, many surface modification strategies to achieve protein-resistant (antifouling) surface have been developed.

Surface chemistry processes for producing antifouling coatings and immobilization of biological recognition elements are pervasive in many commercialized medical applications, including implant biomaterials, chromatography separations, and biosensors. For example, the glucose biosensor is one of the most commercialized biosensor technologies. It is at the heart of the hand-held devices used by millions of diabetic patients for monitoring glucose levels. In a glucose biosensor, an enzyme layer of glucose oxidase immobilized on an oxygen electrode selectively catalyzes the oxidation of glucose in a blood sample. Then, the concentration of the reaction product, H_2O_2, is monitored by either an electrochemical or an optical transducer to detect the glucose concentration in the blood sample. One of major challenges in the development of the new generation of enzyme-based biosensors is to find new enzyme immobilization approaches to maintain the activity of the immobilized enzyme for a long time period.

At the more fundamental level, a detailed understanding of how surface properties control the biological activity at biointerfaces is the key to engineering new biomaterials for advanced biological applications. Advances in surface analysis make it possible to characterize the surface chemistry and structure of increasingly complex biointerfaces. For example, the PEG-terminated self-assembly monolayers have been studied by using surface-specific optical spectroscopy techniques to unveil the molecular structural origin associated with their antifouling properties.

Anticorrosion Technology

In many practical applications, high reactivity of surfaces may cause undesirable chemical (corrosion) or mechanical (wear and friction) processes, which lead to material and energy waste. Depending on the materials (metal, oxide, and polymer) and the surrounding environment (gas, corrosive liquid, temperature, and radiation), there are many surface chemical processes that can lead to the progressive corrosion of material surfaces. On a nonnoble metal surface in an oxidative moist environment, severe corrosion usually occurs through an electrochemical process.

Anticorrosion technology uses various surface chemical processes to coat a material with a protective layer. Most of the protective layers consist of metal oxides (e.g., chromium oxide on stainless steels) or polymers (e.g., Teflon on pans) that are inert in the environment of their specific applications. A protective layer usually serves as a kinetic barrier to slow down corrosion process. The long-term stability of the protective layers is the major concern in many applications. For example, due to corrosion, the aluminized-Teflon coating on the Hubble Space Telescope has been severely degraded after 19 y of exposure to atomic radicals and UV radiation in low Earth orbit.

The design of protective layers with enhanced stability requires surface chemistry investigations on the structures, growth, and breakdown mechanisms of the protective layers. Many surface analysis tools are used to elucidate the chemical composition and morphology of the protective layers, as well as the kinetics of their growth and breakdown in various corrosive environments. These research findings have provided useful information for designing new protective layers with enhanced stability. Several current challenges in this field include the following:

1. The need for detection techniques with high spatial resolution to investigate protective layers that are a few nanometers thick and the corrosion processes that occur locally on the defect sites on the thin layer.

2. The development of model systems that simulate real corrosion processes where the corrosive environment may evolve as the corrosion progresses, or where the harsh corrosive environments are hard to reproduce in laboratories.

Lubricant Technology

A lubricant is a material confined between two hard surfaces in relative motion. Its existence helps reduce the friction and wear at the interfaces. Most of the fluid-base lubricants consist of viscous base stocks (e.g., paraffinic-, aromatic-, and naphthenic-hydrocarbon–type molecules with carbon number ranging from 14 to 40+) and additives to enhance performance under heavy-duty and high temperature conditions. This type of lubricant is widely used in internal combustion engines, vehicle and industrial gear boxes, compressors, turbines, and hydraulic systems. Solid lubricants are layered materials such as graphite, tungsten disulfide, and molybdenum disulfide. These solids are used in high-temperature gas turbines and space applications where fluid-based lubricants are not stable because of vaporization. Other lubricants of great technological importance include ultra-thin lubricant films coated on a computer hard disk surface to prevent wear by the disk head, biocompatible lubricants used in artificial joints, and organic monolayer lubricants in microelectro–mechanical systems.

The terminological introduction of tribology, the science of friction, wear, and lubrication, in the late 1960s reflected the urgent need for scientific focus on molecular mechanisms of lubrication for the development of lubricant technology. In past decades, many tribological studies have focused on properties of thin lubricant layers that hold two sliding surfaces apart. Now, it is known that, under low sliding speeds and/or high loading conditions, a "boundary lubrication" process occurs. In this process, an active lubricant is a thin film only several molecular layers thick. A strongly adsorbed and oriented monolayer is formed on each of the slide surfaces, which protects the surfaces against wear. The shear plane resides within the thin lubricant film, and the low shear strength between lubricant molecular layers reduces the friction.

The development of environmentally benign lubricants is a major driving force for advances in tribology. A fluid-based lubricant is a complex mixture. Detailed understanding of how various components function in the lubrication process may assist the discovery of more benign alternatives. Solid lubricants are usually more environmentally friendly than fluid-based lubricants. However, severe performance degradation in the presence of humidity or oxygen limits the applications for using solid lubricants exposed to the Earth's atmosphere. The surface chemistry of their degradation is an important research subject for improving their chemical stability.

Adsorption

Adsorption is defined as the deposition of molecular species onto the surface. The molecular species that gets adsorbed on the surface is known as adsorbate and the surface on which adsorption occurs is known as adsorbent. Common examples of adsorbents are clay, silica gel, colloids, metals etc.

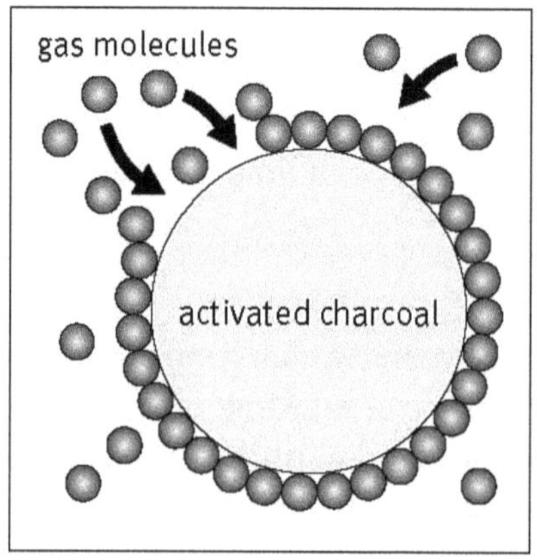

Figure: Adsorption

Adsorption is a surface phenomenon. The process of removal of adsorbent from the surface of adsorbate is known as desorption.

Difference between Absorption and Adsorption

Absorption	Adsorption
Substance penetrates the surface.	Surface phenomenon.
It occurs at uniform rate.	Rate increases initially than it decreases.
It is unaffected by temperature.	It is affected by temperature.
It is an endothermic process.	It is an exothermic process.
It is same throughout the material.	Concentration on the surface of adsorbent is different from that in the bulk.

Mechanism of Adsorption

The amount of heat evolved when one mole of the adsorbate is adsorbed on adsorbent is called enthalpy of adsorption. Adsorption is an exothermic process and enthalpy change is always negative. When adsorbate molecules are adsorbed on the surface, freedom of movement of molecules become restricted and this results in decrease in entropy. Adsorption is a spontaneous process at constant pressure and temperature, thus Gibb's free energy is also decreased.

Types of Adsorption

There are two types of Adsorption – Physical Adsorption or Physiosorption and Chemical Adsorption or Chemisorption.

Physical Adsorption

It involves adsorption of gases on solid surface via weak van der Waal's forces.

Characteristics of Physical Adsorption

- There is no specificity in case of physical adsorption. Every gas is adsorbed on the surface of the solid.
- Nature of the adsorbate. Easily liquefiable gases are strongly adsorbed physically.
- Physical adsorption is reversible in nature. If pressure is increased volume of gas decreases as a result more gas is adsorbed. So, by decreasing the pressure, gas can be removed from the solid surface. Low temperature promotes physical adsorption and high temperature decreases the rate of adsorption.
- More surface area more is the rate of adsorption. Porous substances and finely divided metals are good adsorbents.
- Physical adsorption is an exothermic process.
- No activation energy is needed.

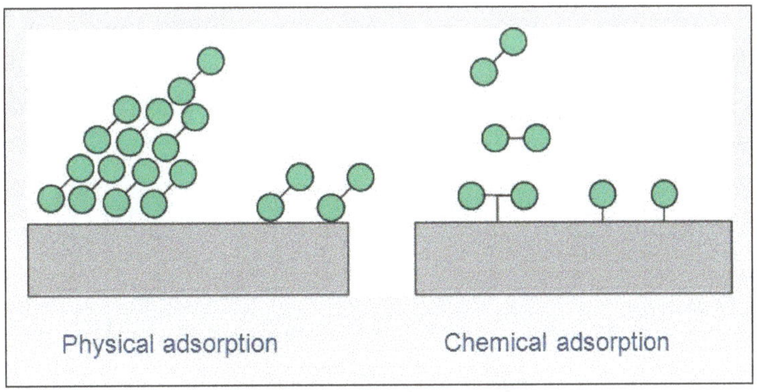

Figure: Types of adsorption

Chemical Adsorption or Chemisorption

When the gas molecules or atoms are held to the solid surface via chemical bonds, this type of adsorption is chemical adsorption or chemisorption.

Characteristics of Chemical Adsorption

- This type of adsorption is specific as compared to physical adsorption. Adsorption occurs only if there is formation of chemical bonds between the adsorbate and adsorbent.
- Chemical adsorption is irreversible. It is an exothermic process but the process occurs slowly at low temperature. Chemisorption is accompanied by increase in temperature. High pressure promotes chemisorption.
- Chemisorption increases with increase in surface area.

- Due to chemical bond formation enthalpy of chemisorption is high.
- Activation energy is needed.
- It results in unimolecular layer.

Adsorption Isotherms

Adsorption isotherm is a graph or a relation between the amounts of adsorbate adsorbed on the surface of adsorbent and pressure at a constant temperature. Different adsorption isotherm was studied by different scientists.

Freundlich Adsorption Isotherm

Freundlich proposed an empirical relationship between amount of gas adsorbed by unit mass of adsorbent and pressure at a particular temperature. Following equation was proposed for freundlich adsorption isotherm,

$$x/m = k.p^{1/n} \left(n > 1\right)$$

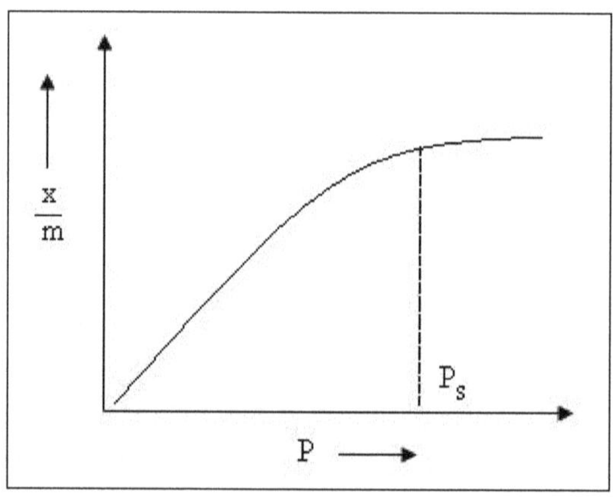

Figure: Adsorption isotherm

- x is the mass of the gas adsorbed.
- m is the mass of the adsorbent.
- p is the pressure.
- *k* and n are constants which depends on the nature of the adsorbent and the gas at a particular temperature.

Taking log of the above equation, the following equation will be observed:

$$\log x/m = \log k + 1/n \log p$$

x/m is plotted on y axis and log p is on x axis. If straight line is observed than only freundlich isotherm is verified.

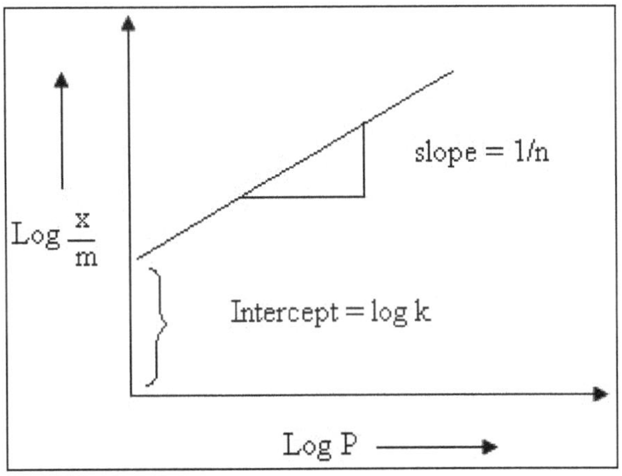

Figure: Freundlich isotherm

Slope gives 1/n and intercept gives log k. The value of 1/n varies from 0 to 1.

If 1/n is 0, adsorption is independent of pressure.

If 1/n is 1, adsorption changes with pressure.

Adsorption from Solution Phase

Solids also adsorb from solutions. For example, when a solution of acetic acid in water is mixed with charcoal, some of the acid is adsorbed by the charcoal.

Characteristics of adsorption from the solution phase:

- Adsorption decreases with increase in temperature.
- More surface area more is the rate of adsorption.
- Adsorption also depends on concentration of solute in a solution.
- Adsorption also depends on nature of adsorbate and nature of adsorbent.

Freundlich explains the adsorption from solution phase using concentration of the solution instead of pressure:

$$x/m = kC^{1/n}$$

Taking log of the above equation, the following reaction will be obtained:

$$\log x/m = \log k + 1/n \log C$$

x/m against log C will give straight line.

Factors Affecting the Adsorption

- Temperature is an important factor that affects the adsorption. Adsorption occurs best at low temperature. As adsorption is an exothermic process, low temperature will derive the forward reaction.

- Adsorption increases with increase in pressure up to certain extent until saturation is reached. After saturation, has achieved no more adsorption will occur irrespective of the pressure applied. The relationship between the extent of adsorption and temperature at any constant pressure is called Adsorption Isobar.
- As adsorption is a surface phenomenon, surface area will increase the rate of adsorption.
- Easily liquefiable gases are easily adsorbed.

Applications of Adsorption

- High vacuum can be created using adsorption strategy. For creating vacuum activated charcoal is used.
- Gas masks used in coal mines are based on adsorption principle. These gas masks are used to adsorb poisonous gases. This makes the air purified for breathing.
- Silica and aluminum gels are used to adsorb moisture to reduce humidity.
- Noble gases can be separated using charcoal as an adsorbent.
- Adsorption of drugs are used to kill germs.
- Chromatographic analysis is based on phenomenon of adsorption.
- Sugar is decolorized by treating sugar solution with charcoal powder. The latter adsorbs the undesirable colors present.
- Adsorption also plays an important role in paint industry. The paint should not contain dissolved gases, as otherwise the paint does not adhere well to the surface to be painted and thus will have a poor covering power.
- This method is also used in the formation of the stable emulsions in cosmetics and syrups.
- The cleaning action of soaps and detergents are also due to adsorption.

Langmuir Adsorption Isotherm

Irving Langmuir proposed another Adsorption Isothermwhich explained the variation of Adsorption with pressure. Based on his theory, he derived Langmuir Equation which depicted a relationship between the number of active sites of the surface undergoing adsorption and pressure.

Assumptions of Langmuir Isotherm

Langmuir proposed his theory by making following assumptions:

- Fixed number of vacant or adsorption sites are available on the surface of solid.
- All the vacant sites are of equal size and shape on the surface of adsorbent.

- Each site can hold maximum of one gaseous molecule and a constant amount of heat energy is released during this process.
- Dynamic equilibrium exists between adsorbed gaseous molecules and the free gaseous molecules.

$$A(g) + B(S) \underset{\text{desorption}}{\overset{\text{Adsorption}}{\rightleftharpoons}} AB,$$

where, A (g) is unadsorbed gaseous molecule, B(s) is unoccupied metal surface and AB is Adsorbed gaseous molecule.

- Adsorption is monolayer or unilayer.

Derivations of the Langmuir Adsorption Equation

Calculation of Equilibrium Constant

Langmuir proposed that dynamic equilibrium exists between adsorbed gaseous molecules and the free gaseous molecules. Using the equilibrium equation, equilibrium constant can be calculated.

$$A(g) + B(S) \underset{K_d}{\overset{K_a}{\rightleftharpoons}} AB$$

Where K_a represents equilibrium constant for forward reaction and K_d represents equilibrium constant for backward direction.

According to Kinetic theory,

Rate of forward reaction = $K_a[A]\ [B]$

Rate of backward reaction = $K_d[AB]$

At equilibrium, Rate of forward reaction is equal to Rate of backward reaction:

$$K_a[A]\ [B] = K_d[AB]$$

$$or, \frac{K_a}{K_d} = \frac{[AB]}{[A][B]}$$

$$k = \frac{K_a}{K_d} = \frac{[AB]}{[A][B]}$$

The above equation represents the equilibrium constant for distribution of adsorbate between the surface and the gas phase.

Derivation

Langmuir Equation which depicts a relationship between the number of active sites of the surface undergoing adsorption (i.e. extent of adsorption) and pressure.

To derive Langmuir Equation and new parameter 'θ ' is introduced. Let θ the number of sites of the surface which are covered with gaseous molecules. Therefore, the fraction of surface which are unoccupied by gaseous molecules will be $(1 - \theta)$.

Now, Rate of forward direction depends upon two factors: Number of sited available on the surface of adsorbent, $(1 - \theta)$ and Pressure, P. Therefore rate of forward reaction is directly proportional to both mentioned factors.

$$\text{Rate of forward reaction} \propto P\,(1 - \theta)$$

$$\text{Rate of adsoption} \propto P\,(1 - \theta)$$

$$\text{Or, Rate of adsoption} = K_a P(1 - \theta)$$

Similarly, Rate of backward reaction or Rate of Desorption depends upon number of sites occupied by the gaseous molecules on the surface of adsorbent.

$$\text{Rate of adsoption} \propto \theta$$

$$\text{Or, Rate of adsoption} = K_d\,\theta$$

At equilibrium, rate of adsorption is equal to rate of desorption:

$$K_a P(1 - \theta) = K_d \theta$$

We can solve the above equation to write it in terms of θ:

$$K_a P - K_a P\,\theta = K_d\,\theta$$

$$K_a P = K_a P\,\theta + K_d\,\theta$$

$$K_a P = (K_d + K_a P)\,\theta$$

$$\theta = \frac{K_a P}{K_d + K_a P}$$

Divide numerator and denominator on RHS by K_d, we get:

$$\theta = \frac{\frac{K_a}{K_d} P}{\frac{K_d}{K_d} + \frac{K_a}{K_d} P}$$

Now put,

$$K = \frac{K_a}{K_d}$$

in above equation we get:

$$\theta = \frac{KP}{1 + KP}$$

This is known as Langmuir Adsorption Equation.

Alternate form of Langmuir Adsorption Equation

Langmuir adsorption equation can be written in an alternate form in terms of volume of gas adsorbed. Let V be volume of gas adsorbed under given sets of conditions of temperature and pressure and V_{mono} be the adsorbed volume of gas at high pressure conditions so as to cover the surface with a unilayer of gaseous molecules.

$$\theta = \frac{V}{V_{\text{mono}}}$$

Substituting the value of θ in Langmuir equation:

$$\frac{V}{V_{\text{mono}}} = \frac{KP}{1+KP}$$

$$\text{or } V_{\text{mono}} = 1 + \frac{1}{KP}$$

Or in terms of pressure P we get:

$$\frac{P}{V} = \frac{P}{V_{\text{mono}}} + \frac{1}{KV_{\text{mono}}}$$

Langmuir Adsorption Equation in alternate form.

Thus, if we plot a graph between P/V vs. P, we will obtain a straight line with:

$$Slope = \frac{1}{V_{mono}} \text{ and } Intercept = \frac{1}{KV_{mono}}$$

Limitations of Langmuir Adsorption Equation

1. The adsorbed gas has to behave ideally in the vapor phase. This condition can be fulfilled at low pressure conditions only. Thus Langmuir Equation is valid under low pressure only.

2. Langmuir Equation assumes that adsorption is monolayer. But, monolayer formation is possible only under low pressure condition. Under high pressure condition the assumption breaks down as gas molecules attract more and more molecules towards each other. *BET theory* proposed by Brunauer, Emmett and Teller explained more realistic multilayer adsorption process.

3. Another assumption was that all the sites on the solid surface are equal in size and shape and have equal affinity for adsorbate molecules i.e. the surface of solid if homogeneous. But we all know that in real solid surfaces are heterogeneous.

4. Langmuir Equation assumed that molecules do not interact with each other. This is impossible as weak force of attraction exists even between molecules of same type.
5. The adsorbed molecules has to be localized i.e. decrease in randomness is zero ($\Delta S = 0$).This is not possible because on adsorption liquefaction of gases taking place, which results into decrease in randomness but the value is not zero.

From above facts we can conclude that, Langmuir equation is valid under low pressure conditions.

Modifications

The modifications try to account for surface roughness, inhomogeneity, and adsorbate-adsorbate interactions.

Freundlich Adsorption Isotherm

The Freundlich isotherm is the most important multisite adsorption isotherm for rough surfaces.

$$\theta_A = \alpha_F\, p^{C_F}$$

where α_F and C_F are fitting parameters. This equation implies that if one makes a log-log plot of adsorption data, the data will fit a straight line. The Freundlich isotherm has two parameters while Langmuir's equations has only one: as a result, it often fits the data on rough surfaces better than the Langmuir's equations.

A related equation is the Toth equation. Rearranging the Langmuir equation, one can obtain:

$$\theta_A = \frac{p_A}{\frac{1}{K_{eq}^A} + p_A}$$

Toth modified this equation by adding two parameters, α_{To} and C_{To} to formulate the Toth equation:

$$\theta^{C_{T_0}} = \frac{\alpha_{T_0}\, p_A^{C_{T_0}}}{\frac{1}{K_{eq}^A} + p_A^{C_{T_0}}}$$

Temkin Adsorption Isotherm

This isotherm takes into account indirect adsorbate-adsorbate interactions on adsorption isotherms. Temkin noted experimentally that heats of adsorption would more often decrease than increase with increasing coverage.

The heat of adsorption ΔH_{ad} is defined as:

$$\frac{[A_{ad}]}{p_A[S]} = K_{eq}^A \propto \mathrm{e}^{-\Delta G_{ad}/RT} = \mathrm{e}^{\Delta S_{ad}/R}\,\mathrm{e}^{-\Delta H_{ad}/RT}$$

He derived a model assuming that as the surface is loaded up with adsorbate, the heat of adsorption of all the molecules in the layer would decrease linearly with coverage due to adsorbate-adsorbate interactions:

$$\Delta H_{ad} = \Delta H_{ad}^{0}(1-\alpha_T \theta)$$

where α_T is a fitting parameter. Assuming the Langmuir Adsorption isotherm still applied to the adsorbed layer, K_{eq}^{A} is expected to vary with coverage, as follows:

$$K_{eq}^{A} = K_{eq}^{A,0} e^{(\Delta H_{ad}^{0}(1-\alpha_T \theta)/kT)}$$

Langmuir's isotherm can be rearranged to this form:

$$K_{eq}^{A} p_A = \frac{\theta}{1-\theta}$$

Substituting the expression of the equilibrium constant and taking the natural logarithm:

$$\ln(K_{eq}^{A,0} p_A) = \frac{-\Delta H_{ad}^{0} \alpha_T \theta}{kT} + \ln\left(\frac{\theta}{1-\theta}\right)$$

BET Equation

Brunauer's model of multilayer adsorption, that is, a random distribution of sites covered by one, two, three, etc., adsorbate molecules.

Brunauer, Emmett and Teller derived the first isotherm for multilayer adsorption. It assumes a random distribution of sites that are empty or that are covered with by one monolayer, two layers and so on, as illustrated alongside. The main equation of this model is:

$$\frac{[A]}{S_0} = \frac{c_B x_B}{(1-x_B)[1+(c_B-1)x_B]}$$

where,

$$x_B = p_A K_m, \qquad c_B = \frac{K_1}{K_m}$$

and *[A]* is the total concentration of molecules on the surface, given by:

$$[A]=\sum_{i=1}^{\infty} i[A]_i=\sum_{i=1}^{\infty} iK_1K_m^{i-1}p_A^i[A]_0$$

where

$$K_i=\frac{[A]_i}{p_A[A]_{i-1}}$$

in which $[A]_o$ is the number of bare sites, and $[A]_i$ is the number of surface sites covered by *i* molecules.

Adsorption of a Binary Liquid on a Solid

This topic describes the surface coverage when the adsorbate is in liquid phase and is a binary mixture.

For ideal both phases - no lateral interactions, homogeneous surface - the composition of a surface phase for a binary liquid system in contact with solid surface is given by a classic Everett isotherm equation (being a simple analogue of Langmuir equation), where the components are interchangeable (i.e. "1" may be exchanged to "2") without change of equation form:

$$x_1^s=\frac{Kx_1^l}{1+(K-1)x_1^l}$$

where the normal definition of multicomponent system is valid as follows:

$$\sum_{i=1}^{k} x_i^s=1\sum_{i=1}^{k} x_i^l=1$$

By simple rearrangement, we get:

$$x_1^s=\frac{K[x_1^l/(1-x_1^l)]}{1+K[x_1^l/(1-x_1^l)]}$$

This equation describes competition of components "1" and "2".

Molecular Diffusion at Surfaces

Surface diffusion (SD) is a really common phenomenon playing a highly important part in the field of science and technology. The effect of SD is to move surface atoms, molecules and clusters and allow them to assemble into some desirable configurations or, vise versa, to destroy the configurations that have been purposely created. There exists many processes where surface diffusion has a significant role to play as the growth of crystals and thin films; catalysis; sintering and powder metallurgy; capillary phenomena; corrosion; nanotechnologies of all kinds; strengthening of materials; soldering; a multitude of processes that occur on various biological interfaces, etc.

Two types of surface diffusion can play important roles in coverage and deposition layers; bulk diffusion, which is the global macro-motion of the material within the deposited layer, and surface diffusion, which relates to the motion of metal boundaries.

Surface diffusion is a many body or collective phenomenon. Even in the case when we are able to trace the activities of a single atom, one should realize the effect of its motion or activity is solely because of its continuous interaction with the substrate atoms.

In the figure given below the atom receives fluctuatively some excessive energy from the substrate. This enables it to jump from one residence site to another. Second, the jump ends as soon as the excessive energy is given back from the diffusing particle to the substrate (the heat bath).

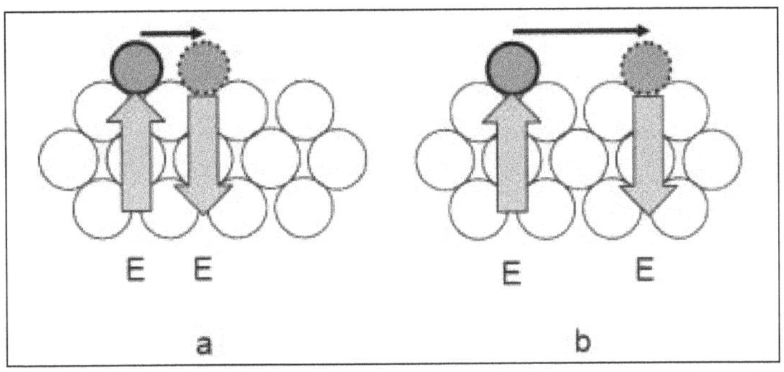

Figure: a) Elementary and b) long jumps of the adsorbed atom on a substrate.
E is the corresponding excess energy which the adatom received and given back to the substrate.

Factors Responsible for Surface Diffusion

One can attribute the surface diffusion phenomenon to the various lateral interactions developed on the surface. The Van der Waals, direct exchange, indirect exchange and dipole-dipole and higher multipole interactions which are responsible for molecules diffusion at surfaces. These interactions vary in both in their intensity as well as nature. The energies associated in the interactions mentioned above are a function of distance between the ad atoms and substrate.

Van der Waals interaction is attractive and its energy decreases with distance as r^{-6} .The energy of electronic indirect interaction oscillates with distance as the wave functions of interacting adatoms overlap in phase or antiphase. Another kind of the indirect interaction is connected with elastic strains induced in the substrate by the adsorbed layer. The strains arise due to lattice misfit between the substrate and the adlayer. The dipole–dipole interaction, which arises because the adsorption bond may have a considerable dipole moment, decays with distance rather slowly (ar^{-3}) and plays an important role in many surface phenomenon.

One can conclude with the available information in regard to the various interactions responsible for surface diffusion that:

- SD is basically a many-body process.
- The kinetics of SD cannot be understood without proper account for interactions between the diffusing particles.

- SD should be investigated in connection with surface phase transitions.

Kinetics of Surface Diffusion

The diffusion coefficient D is defined as the proportionality constant relating the diffusion flux J to the concentration gradient.

$J = -D \dfrac{\partial n}{\partial x}$, where we have assumed concentration n depends only on x.

For diffusion of non-interacting species D is just the product of the mean-square jump distance λ^2 and $\Gamma(0)$, the jump frequency of adatoms along the x-direction,

$$D = \Gamma(0)\lambda^2$$

Substituting known expressions for $\Gamma(0)$ and λ^2, we get the following expression for the diffusion coefficient,

$$D = D_0 \exp\left(-E_d / kT\right)$$

With the pre-exponential factor given by:

$$D_0 = \nu \exp\left(\Delta S^{\neq} / k\right)\lambda^2$$

With the aid of Fick's second law (continuity equation) the diffusion coefficient can be evaluated from the experimental data measured in a concentration gradient:

$$\frac{\partial n}{\partial t} = -\frac{\partial J}{\partial x} = \frac{\partial}{\partial x}\left(D\frac{\partial n}{\partial x}\right)$$

If *D* depends on x neither explicitly nor via concentration, the latter equation is reduced to:

$$\frac{\partial n}{\partial t} = D\frac{\partial^2 n}{\partial x^2}$$

Effect of Lateral Interactions on SD Parameters: Thermodynamic Factor

In the presence of lateral interactions an additional driving force of diffusion acts together with the concentration gradient. Due to deformations of the potential relief, the frequencies of adatom jumps in opposite directions become unequal. In this case the diffusion flux must be written in a more general form:

$$J = -L\frac{\partial \mu}{\partial x},$$

where *L* is a phenomenological transport coefficient and μ is the chemical potential of adatoms. This relation may be put down as follows:

$$J = -L\frac{\partial \mu}{\partial n}\frac{\partial n}{\partial x}$$

which allows one to introduce the effective diffusion coefficient:

$$D = L\frac{\partial \mu}{\partial n}$$

In its turn, the transport coefficient is given by:

$$L = Bn$$

where B is the mobility of adatoms which can be related through formula to the self-diffusion coefficient D_i the latter characterizes of adatoms in a homogeneous layer:

$$B = D_i/kT$$

Thus, we have:

$$D(n) = \frac{D_i(n)}{kT} n \frac{\partial \mu}{\partial n} = D_i(n)\frac{\partial(\mu / kT)}{\partial \ln n}$$

The term $\frac{\partial(\mu / kT)}{\partial \ln n}$ is called the thermodynamic factor (TDF).

At low concentrations $(\theta \ll 1)$ and in the absence of interaction, μ is given by:

$$\mu = \mu^\circ + kT \ln \theta$$

where μ° is the standard chemical potential.

In this case, as one should expect, $\partial\left(\mu / kT\right) / \partial \ln \theta = 1$

In Langmuir's layer the interaction of adatoms is also absent, but the restriction $\theta \ll 1$ is not imposed and each adsorption site is assumed to be occupied by a single atom. For this model,

$$\mu = \mu^\circ + kT \ln \frac{\theta}{1-\theta}$$

For interacting adatoms μ can be approximated as:

$$\mu = \mu^\circ + kT \ln \frac{\theta}{1-\theta} - q(\theta)$$

where $q(\theta)$ is the coverage dependent adsorption heat; so,

$$\frac{\partial\left(\mu / kT\right)}{\partial \ln \theta} = \frac{1}{1-\theta} - \frac{\partial\left(q / kT\right)}{\partial \ln \theta}$$

The TDF can also be calculated using experimental data on adsorption isotherms. In particular, if the pressure of the gas in equilibrium with an adlayer is sufficiently low so that ideal gas approximation holds, one obtains:

$$\frac{\partial(\mu/kT)}{\partial \ln \theta} = \frac{\partial \ln p}{\partial \ln \theta}$$

It is quite evident that lateral interaction between adatoms affects not only the absolute value of D, but also its temperature dependence. In general, the effective activation energy of surface diffusion in a concentration gradient, which we denote by E_d as before, is not equal to the activation energy of adsorbate self-diffusion E_{di}.

$$E_d = E_{di} + \frac{\partial \ln p}{\partial \ln \theta} \frac{\partial q}{\partial \ln \theta}$$

Here, as given previously, p is the pressure of the gas in equilibrium with the adlayer at a coverage θ. Both derivatives are calculated at T = const. It goes without saying that the lateral interaction changes the value of the pre-exponential factor as well.

The above relations account for the origin of the interrelation between surface diffusion and phase transitions in adsorbed layers; on the one hand, the variation of μ with coverage manifests itself in changes of D and other diffusion parameters and, on the other hand, this variation governs the whole pattern of phase transitions in the adlayer. It should be pointed out, however, that the functional relation between D and $\frac{\partial(\mu/kT)}{\partial \ln \theta}$ is complicated by the coverage dependence of mobility of adatoms (and that of Di, respectively), because in different phases both the diffusion parameters and even the mechanisms may be different. Furthermore, this problem cannot be solved in terms of a purely thermodynamic treatment; for this purpose one must necessarily proceed to a detailed analysis of proper atomic models.

Surface Diffusion of Carbon on Fe

Carbon dissolved in transition metals is one of the most important nonmetallic impurities. Carbon is believed to be an interstitial element in cubic lattices and to diffuse through the interstitial sites because it is much smaller than host-metal atoms. We are going to see the properties of carbon diffused on alpha iron. It has been experimentally observed that carbon segregated to metal surfaces is displaced by sulfur at high temperature as expected because the segregation enthalpy of carbon is smaller than that of sulfur. In contrast to Sulfur, carbon is known to enhance the bonding at grain boundaries. Sulfur forms ionic bonds and draws the charge from the metal atoms, leading to weakening of the metalmetal bonds. But carbon forms a covalent Fe 3d-C2s,2p interactions with iron there by gets more stabilized.

The ionization potential input parameters for the C 2s and 2p orbitals are decreased by1.5 eV from their atomic energy levels because of charge transfer to C. When a C atom is introduced into an interstitial site, not only does its interaction with neighbouring Fe atoms affect the energy but also their relaxation around it. Therefore, the energy change, ΔE, can be divided into two parts: ΔE_{Fe}, due to Fe atom relaxation and ΔEc due to Fe-C bonding. Thus $\Delta E = \Delta E_{Fe} + \Delta Ec$. ΔE_{Fe} should be positive since Fe atoms move away from their equilibrium position toward surrounding Fe atoms of the lattice and ΔEc is expected to be negative since C binds to its neighboring Fe atoms. ΔE_{Fe} is determined from the comparison of un-relaxed and relaxed systems without C atoms, and ΔEc comes from subtracting ΔE_{Fe} from ΔE . Since in the ASED-MO theory the total energy is the sum of the two-body repulsion energy, ER, and the molecular-orbital stabilization energy, EM, the following formulas can be written: $\Delta E = \Delta E_R + \Delta E_{MO}$, $\Delta E_{Fe} = \Delta E_{R,Fe} + \Delta E_{MO,Fe,}$ $\Delta E_c = \Delta E_{R,C} + \Delta E_{MO,C}$. It is assumed that the change in spin polarization of the Fe atoms s induced only by coordination of C to neighboring Fe atoms, not by the relaxation of Fe atoms.

A low-energy electron-diffraction (LEED) study has revealed that C adsorbs at the fourfold sites on Fe(100). The large four-layer Fe_{42}(100) and the smallest Fe_5(100) cluster models were used to study the adsorption of C on the (100) surface. The height of C from the Fe_{42}(100) surface layer is calculated to be 0.64 A. No experimental value is available but it can bc compared with the height of O, which is 0.53+-0.06 A on the Fe(100) surface obtained from LEED analysis. The distance between C and the second-layer Fe directly below is 2.07 A, which is shorter than 2.13 A between C and the surface layer Fe atoms.

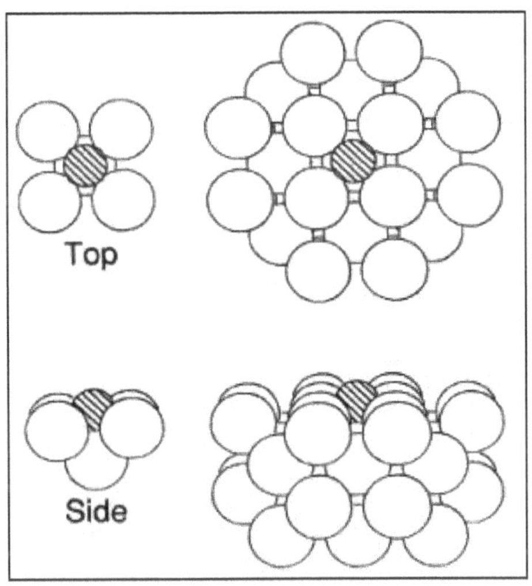

The binding energy (BE) of C on the Fe_{42}(100) surface is calculated to be 7.89 eV. This is close to the value of 7.56 eV deduced from the segregation enthalpy and the relative partial enthalpy of carbon in αiron. No direct measurement is available. When the same height of C is used on the Fe_5(100) surface, the BE is calculated to be 7.23 eV. The molecular-orbital diagram for the large $Fe_{42}(100)C_{ad}$ cluster model is expected to be close to the bulk limit for isolated C atom adsorption. For the MO analysis, only the $Fe_5(100)C_{ad}$ cluster is used and the MO diagram for a-symmetry orbital interactions is shown in figure. The C 2pz orbital interacts mainly with Fe 3d orbitals and stabilizes them. The C 2s orbital undergoes a bonding stabilization by mixing with Fe 4s +3d orbitals but also pushes up the stabilized $C\,2p_z + Fe\ 3d$ bonding orbitals by antibonding mixing with the C 2pz orbital. When the interaction of the C 2s orbital with Fe orbitals is removed by moving down the C 2s level to -50 eV and shrinking the orbital by using a Slater exponent of 30 a.u., the BE of C on the Fe_{42}(100) surface becomes 7.79 eV, which is 0.10 eV, less than when the C 2s orbital interactions is permitted. Consequently, the C 2s orbital interaction with the Fe surface band is bonding, unlike in the S case, though very weak.

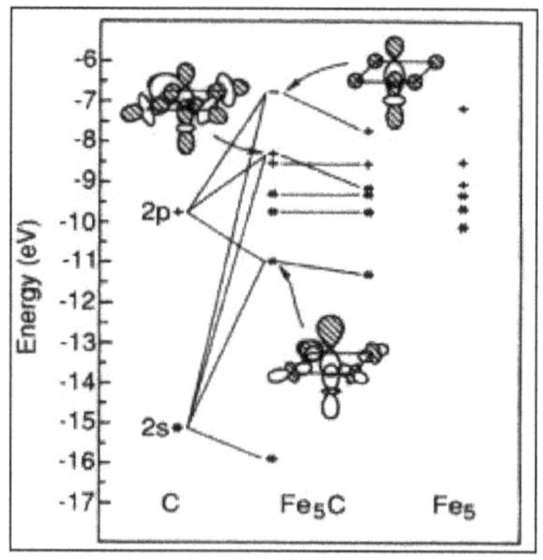

C at interstitial Sites in a-iron

In bcc a-iron the interstitial C atom has been believed to occupy either an octahedral (O) site or a tetrahedral (T) site and diffuse from one site to the other. The O site has two Fe $toms at a distance of 1.43 A and four Fe atoms 2.02 A away from the center. An Fe, cluster model was used for studying C at interstitial site. This enables relaxations of surrounding Fe atoms to be included. The five sites for an interstitial C atom are o, T, one-fourth (O-M), one-half or middle (M), and three fourths (M-T) of the distance between the O and T sites. The optimized structures for C at the o, T, and o-M sites are shown in figure. In the O structure, the two nearest-neighboring Fe atoms move away 0.46 A and the next-nearest four Fe atoms move 0.11 A away from their original lattice points. The distance between C and the nearest Fe atoms is 1.89 A and between C and the next nearest ones it is 2.13 A. In the T structure, four Fe atom adjacent to C expand by 0.37 A to a distance of 1.97 A. Twenty more Fe atoms are relaxed. Eight of them move away 0.18 A from their original lattice points and the rest of them move 0.04 A. When C occupies the O-M site, the two Fe atoms at apexes of the O structure expand 0.47 A to a distance of 1.91 A. The two Fe atoms close to C at corners of the square move 0.18 A away to a distance of 2.09 A and the other two Fe atoms at corners of the other side also relax 0.12 A away.

The four Fe atom surrounding the two Fe atoms at the apexes move 0.18 A. The other four atoms surrounding the two Fe atoms close to C move 0. 13 A. Twelve more surrounding Fe atoms relax < 0.07 A. The change in energy for C moving from the O site through the T site is shown in the figure. The o and T sites are comparable in energy to each other but both of them are predicted to be less stable than the O-M site. It was found that the sum of $\Delta E_{MO,C}$ and $\Delta E_{R,C}$ is minimum at the O-M site. Also at this site the destabilization due to relaxations of Fe atoms is the least; it is greatest at the O site. As a result, the O-M site turns out to be the most favorable for an interstitial C atom. No conspicuous change in MO levels for the different sites of C can be found except for the C 2p +Fe d bonding orbitals. At the T site these levels are less stabilized compared to those at the other two sites. The C 2s orbital is greatly stabilized and splits far from its empty antibonding counterparts because of the large overlap resulting from the short Fe-C distance.

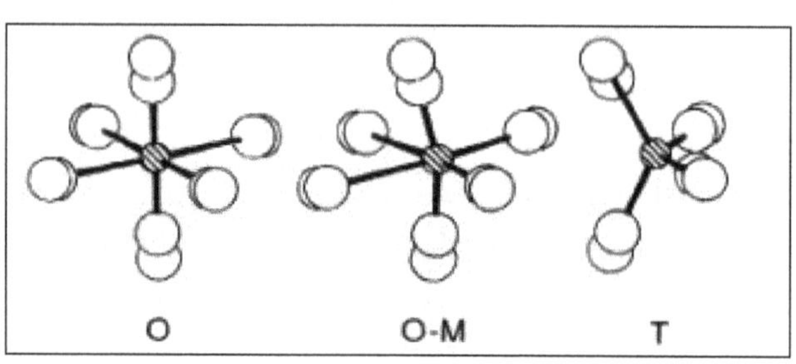

The bonding of a C atom to Fe atoms is achieved predominantly through the overlap of C 2p orbitals with Fe 4s +3d band orbitals irrespective of the position of C in the Fe matrix. The MO energy gain is about 1 eV greater in interstitial sites than on the (100) surface. The result of the C 2s orbital interaction with the Fe 4s +3d band depends on the position of C; it is almost nonbonding on the surface and weakly bonding at interstitial sites. Since the C 2s orbital induces net bonding character, though very weak on the surface and at interstitial sites of a-iron, there is no closed-shell interaction, as was observed for sulfur. Rather, the interaction of a C atom with surrounding Fe atoms is stronger at interstitial sites than on a surface. Therefore, the driving force for C segregation is not the same as for S segregation. We can notice that ΔE for C on the (100) surface is about the same as ΔEc for C at interstitial sites. The difference between two states amounts to ΔE_{Fe}, 0.96 eV, which is the destabilization energy of Fe atoms due to the relaxations. Therefore, the driving force for C segregation comes from the Fe-Fe repulsion energy.

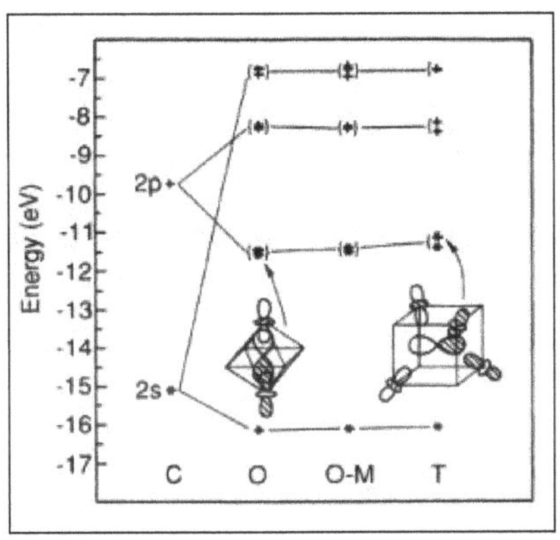

The strong bonding of C to Fe is achieved through the interaction of C 2p orbitals with Fe 4s + 3d band orbitals. The interaction of the C 2s orbital with the Fe 4s +3d band is dependent on the position of C in a-iron. On the (100) surface it shows weak bonding and at the interstitial sites slightly stronger bonding. This is in contrast to substitutional sulfur, for which our past study found a significant closed-shell repulsive interaction between the S valence 3s orbital and the Fe valence band. For an interstitial C atom in a-iron, it is predicted that the octahedral site is the least favorable and the offoctahedral site one-fourth towards a tetrahedral site is the most favorable. It is deduced from the comparison of energy components of C on the (100) surface and in the interstitial site that C atoms segregate to the surface because of the repulsive compression of Fe-Fe bonds surrounding the interstitial C.

Experimental Techniques used in Surface Diffusion Studies

Existing techniques that detect surface diffusion processes can be grouped into two main categories. In the first of them experimentalists observe the processes of equilibration realized via surface diffusion in various nonequilibrium systems (nonequilibrium methods). In the second category of the methods, the systems are in equilibrium.There is no net mass transport on the macroscopic scale in this case, but the particles perform thermal fluctuations around the equilibrium state. These fluctuations are the source of information about the mobility of the particles in such "equilibrium methods". It is understood that each class of the methods gives evidence on particle mobilities in conditions that can be substantially different. This generally results in difference of the values of the diffusion parameters extracted from the equilibrium and nonequilibrium measurements.

Nonequilibrium Methods

The task of such experiments is to record the kinetics of the transition, executed by surface diffusion, from a nonequilibrium state of the object to the equilibrium one. Consider the case of surface self-diffusion. A typical example is the change of the shape of a crystal observed under annealing. Suppose the annealing temperature is low enough to preclude evaporation of the crystal and volume diffusion within it. Then the equilibration, i.e., the minimization of the surface free energy by a shape change, is attained through surface self-diffusion only. For instance, any sharp tip is getting blunter under annealing. Such experiments can be carried out with atomic resolution by field ion microscopy. In some other cases electron microscopy and optical interferometry are also used. The disadvantage of such methods is that their results relate to the surfaces whose atomic structure is changing during the experiment.

The nonequilibrium methods developed to investigate surface diffusion of adsorbates use one or another physical quantity sensitive to adparticle concentration (coverage) on the surface. The measurement of this quantity with an appropriate spatial resolution allows recording the time variation of the adsorbate surface distribution (coverage profile) as the adsorbed layer evolves from a Non-equilibrium initial state to its equilibrium final state.The process of evolution of the coverage profiles can be followed by scanning the surface with such probes as electron, ion and light beams, or a tip in various scanning probe microscopies. The field emission and photoemission microscopies also are successfully employed to display the adsorbate distributions on the substrates.

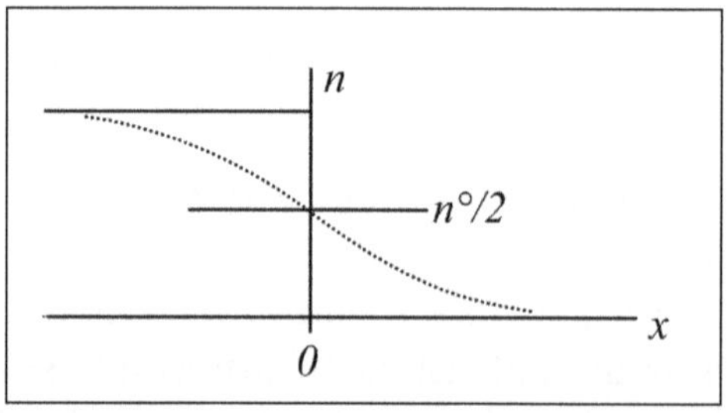

Figure: Concentration profile (dashed line) in the case of diffusion of noninteracting particles from a step-like initial deposit.

Field-ion and scanning tunneling microscopies provide direct views of individual atoms and molecules on surfaces. Other techniques give data, such as electron or ion currents, work function, intensity of atom or light beams, etc., that must be converted to coverage degree (adsorbate surface concentration). This is accomplished through corresponding calibration procedures that establish the relationship between the adatom concentration and the physical values immediately measured in concrete experiments.

The most important features of the experimental techniques applied to study surface diffusion are their spatial resolution and accuracy of coverage determination.

Spacial Resolution

- Field-ion and scanning tunneling microscopies: 10^{-8} cm.
- Photoelectron and other types of electron microscopies and scanning contact-potential microscopy: 1–10 μm.

Scanning contact-potential microscopy provides an accuracy in coverage determination to about 10^{-2} of a monolayer in the coverage intervals where the work function strongly depends on coverage as shown in figure below:

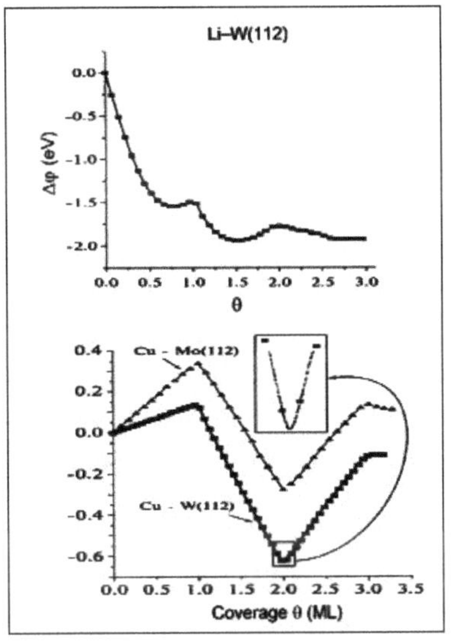

Field Ion Microscopy (FIM)

- Atomic resolution possible.
- Very efficient in following surface diffusion of single atoms and small clusters (oligomers).
- Only applicable to materials which can withstand high electric fields of 10^8 v/cm.
- The small size of the crystal planes on sharp tips used in field ion microscopes makes impracticable quantitative recording of the evolution of coverage profiles and investigations of phase effects in surface diffusion.

Surface Tunneling Microscopy

Visualization possible- But rather high electric fields under the tip and the high-density tunneling current can sometimes induce a specific mobility of the ad particles.

The much more modest resolution of such methods as photoelectron microscopy and scanning contact-potential microscopy (1–10 mm) is rewarded with a high accuracy of coverage determination, non-destructive character of the measurements and the possibility to observe phase effects in surface diffusion over rather large distances.

Equilibrium Methods

Field Ion Microscopy

In FIM, a sharp (< 50 nm tip radius) metal tip is produced and placed in an ultrahigh vacuum chamber, which is backfilled with an imaging gas such as helium or neon. The tip is cooled to cryogenic temperatures (20–100 K). A positive voltage of 5000 to 10 000 volts is applied to the tip. Gas atoms adsorbed on the tip are ionized by the strong electric field in the vicinity of the tip (thus, "field ionization"), becoming positively charged and being repelled from the tip. The curvature of the surface near the tip causes a natural magnification — ions are repelled in a direction roughly perpendicular to the surface (a "point projection" effect). A detector is placed so as to collect these repelled ions; the image formed from all the collected ions can be of sufficient resolution to image individual atoms on the tip surface.

Field emission microscopy (FEM) and STM are two versions of method of surface diffusion investigation based on recording thermal density fluctuations in adsorbed layers.

FEM Fluctuation Technique

FEM consists of a sharp needle emitter and a detector, such as a fluorescent screen. A negative electric field is applied to the emitter, emitting electrons from the surface. An image is formed at the detector due to the different current densities, which originates from the difference in electric fields and work functions on the emitter surface by the Fowler-Nordheim equation.

The original physical quantity recorded in this technique is the field emission current collected from a small (≤100A° in size) surface area. The current fluctuations stem mainly from the fluctuations of the number of adatoms within the probed area, which cause local work function variations. In turn, the density fluctuations in the adlayer can only occur thanks to the existence of a mobility of adatoms. Thus, the investigation of the process of building-up and decay of the adatom density fluctuations gives a possibility to extract the diffusion coefficient from the autocorrelation function of the fluctuations.

STM Fluctuation Technique

In this method, one records the fluctuations of the tunneling current caused by passing the diffusing particles under an immobile tip.

Inelastic Helium Atom Scattering

Over two decades ago, inelastic helium atom scattering (HAS) was developed into a high resolution method to probe the low frequency dynamics of both clean and adsorbate covered surfaces. High resolution helium atom scattering (HAS) has proved to be a useful probe of the low frequency (< 20 meV, <160 cm^{-1}) dynamics of adsorbates on metal surfaces including diffusion.

Thermal energy of helium atoms (E_i=4–100 meV, 30–800 cm^{-1}) have the following advantage over other probe particles, such as electrons and photons.

a) They do not penetrate the surface but, instead, turn away 3–4 Å from the surface where the electron density is low (10^{-4} eÅ^{-3}).

b) Can be generated with very narrow velocity distributions using supersonic expansion, typically $\Delta v / v 1\%$ permitting resolutions in the low sub-meV range. This resolution, combined with the large cross-section for scattering from molecules, makes HAS a very useful and, in some cases, the only probe of low frequency surface dynamics.

c) Low adsorption energy of helium on surfaces coupled with the chemical inertness of helium atoms ensures that the beam itself does not alter the surface.

The HAS Experiment

Diagram showing the essential details of a fixed total scattering angle HAS apparatus. Helium expanding from the high pressure nozzle is collimated and chopped into short bursts. The helium pulses scattered from the surface through a fixed total scattering angle θ_{SD} are detected by a helium specific detector in a TOF mode. The flight times can then be converted into energy transfer to the surface during the collision. The momentum transfer parallel to the surface ΔK is varied by changing the incident angle of the helium atom beam to the surface.

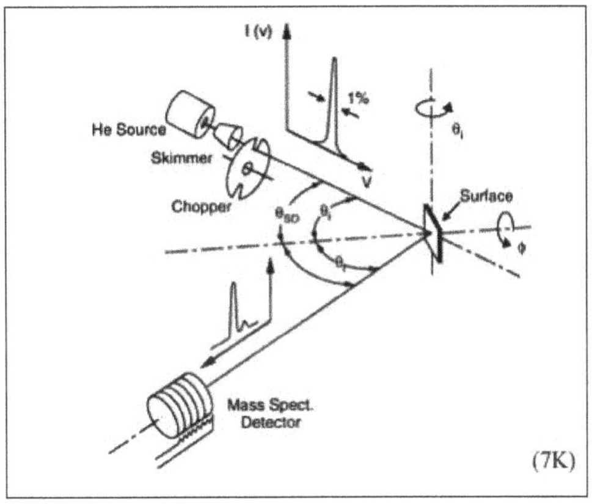

Decay-detected nuclear magnetic resonance.

Effect of Defects on Surface

In order to gain insight into essential physics of this impact of defects on surface diffusion, experimentalists must work with surfaces that possess controllable (intentionally created and well characterized) defect.

The point-like defects such as vacancies, single impurity atoms, etc. act as traps (attracting sites) or antitraps (repulsing sites) for diffusing particles. A trap can be filled by a diffusing atom which then may either attract or repulse the other diffusing atoms. In the former case the trap acts as a nucleation center if the temperature and the adatom density correspond to the region of a first-order phase transition in the diffusing overlayer. In contrast, if the point-like defect (either empty or filled) is a repulsive center, the other diffusing atoms will bypass it.

In the particular case when surface potential corrugation is strongly anisotropic and the diffusion predominantly proceeds along the atomic troughs (channels), the point-like defects situated within the channels can dramatically block the diffusion. A similar strong suppression of surface diffusion by point-like defects is observed in the case of soliton diffusion mechanism. This mechanism has a pronounced collective character: the mass transport is affected by solitons, which represent linear many-particle objects (domain walls) and can be effectively pinned by point-like defects. Linear surface defects such as atomic steps and grain boundaries can have a strong impact on surface diffusion, because the diffusing particles cannot bypass them.

Broadly speaking, diffusion on surface with defects presents a particular case of diffusion in (for our situation, on) disordered media. This is a complicated theoretical problem which is approached by involving such concepts as a certain hierarchy of defects according to their scale, percolation; possible fractal structure of the medium, etc. The diffusion in (on) disordered media differs by some unusual features. For example, the mean-square displacement of the diffusing particles increases in this case nonlinearly versus time. This gave grounds to term such diffusion anomalous in contrast to normal diffusion where (on defect-free surfaces) one has $< x^2 > \alpha\ t$.

Applications of Surface Diffusion

Applications in Nano-Technology

The surface diffusion of large organic molecules on solid surfaces has attracted considerable attention in recent years, partly from the point of view of basic science, partly because of the potential applications in emerging nano technological fields, such as molecular electronics, optoelectronics or nano mechanical sensors. In general, molecular assembly will be controlled both by thermodynamic factors—the molecules tend to assemble on surfaces into the energetically most stable structures; and by kinetic factors—the molecules diffuse and interact at surfaces, thereby creating metastable molecular nanostructures that may hinder the molecules at a given temperature from attaining the minimum energy arrangement. A Nano scale understanding of the complex dynamics of large molecules at surfaces is essential for the bottom-up design of molecular nanostructures. It has been shown that the diffusion coefficient of the complex organic molecule known as Violet Lander (VL, C108H104) on Cu(110) can be changed by two orders of magnitude by using the STM at low temperatures to switch between two adsorption configurations that differ only in the molecular orientation with respect to the substrate lattice.

Otero et al. made a remarkable breakthrough a few years back when they showed that orientation and shape of large and complex organic molecules influence their surface diffusion dynamics. They showed that the diffusion coefficient of the complex organic molecule known as the Violet Lander (VL, C108H104) on a Cu (110) surface can be changed by two orders of magnitude by deliberately manipulating the VL molecules to change their orientation with respect to the substrate. Combining the results with molecular dynamics simulations, we show that the experimental findings can be attributed to the different registries between the molecule and the substrate for the different molecular orientations. An orientation dependence of the molecular diffusion coefficient plays a significant role in controlling the molecular assembly of large organic molecules whose shape allows them to adopt different adsorption orientations and, thus, different registries with the substrate.

The Violet Lander molecule consists of a polyaromatic π-system (molecular board) with four 3, 5-di-t-butylphenyl substituents σ-bonded to the central board; the so-called 'spacer legs'. On deposition onto a Cu(110) surface the molecules appear in STM images as entities with four protrusions corresponding to high electron tunneling probability through the spacer legs into the substrate To explore the energy landscape for the VL molecule on the Cu(110) surface Otero et al, used the STM tip as a tool to push the molecule gently in a direction perpendicular to the board of the molecules, thereby manipulating the orientation of individual molecules at low temperatures (160–200 K). This procedure results in a rotation of the molecular board by 70° with respect to the close packed direction of the Cu substrate in about 60% of the cases.

When the molecules are manipulated on a flat terrace they can be forced into no other orientations. The rotated configuration can only be observed at temperatures below 200 K, and in this range of temperatures they do not spontaneously switch to the non-rotated configuration. They are thus trapped in a metastable state that is kinetically stabilized at low temperatures. No significant conformational change of the molecules was found to occur during the manipulation process.

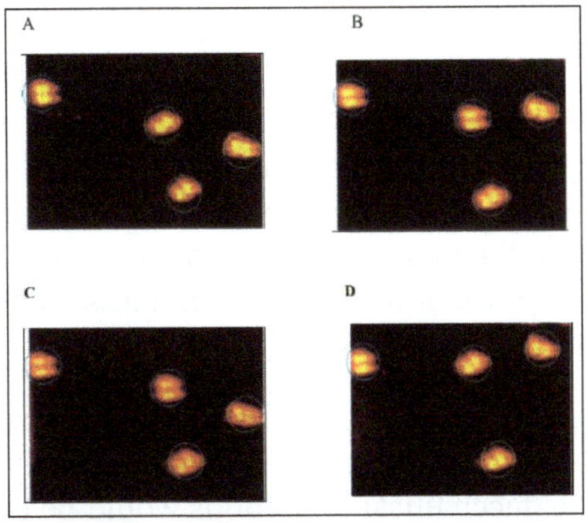

Snapshots of STM movie recorded at 180 K.

The field of view is 28 × 20 m². Before the movie is started, two molecules have been rotated with the STM tip (green circles). Another two molecules are aligned along [11–0] (blue circles). Between A and B only the rotated molecules diffuse, whereas the non-rotated molecules remain in place. However, when molecule 1 is rotated 70°, as happens between the stills B and C, it also starts diffusing (D).

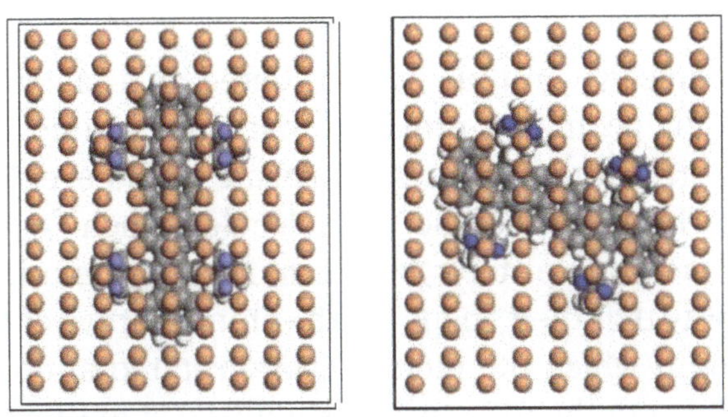

Figure: Bottom views of the adsorbed VL molecule and first Cu layer, showing the different registries between the contact H atoms and the fourfold hollow site for the non-rotated and rotated molecules.

From the STM movies the molecular diffusion coefficient can be determined by tracking the position of the individual molecules. Surprisingly it was found that, whereas the manipulated molecules in the rotated configuration are found to diffuse along the direction with a diffusion coefficient of $D = (4.8 \pm 0.5) \times 10^{-17}$ cm² s⁻¹, the molecules in the parallel configuration do not change their relative positions in the time span of the STM movies (several minutes), corresponding to a diffusion coefficient of less than 5×10^{-19} cm² s⁻¹. They thus concluded that the diffusion coefficient increases by at least two orders of magnitude when the molecular orientation changes from being aligned parallel to the closed-packed direction of the Cu (110) surface to being rotated by 70°.

Although the molecule was immobile before the manipulation, once its board is rotated, it is observed to diffuse with the same diffusion constant as other rotated molecules. Similarly, the molecular diffusion can be stopped at any time simply by flipping the molecule back to its original parallel configuration. The diffusion of the individual VL molecules can thus be activated or deactivated by STM-induced molecular rotations with respect to the close-packed direction of the Cu (110) surface.

These results can be intuitively understood by considering the extent to which the 3D VL shape fits the atomic lattice of the Cu (110) substrate. The VL molecule adsorbed on the Cu substrate exposes eight hydrogen atoms at the bottom of the legs that provide the closest contact between the surface and the molecule. In the non-rotated geometry, these H atoms are in perfect registry with the fourfold hollow sites of Cu (110): the molecular shape fits the atomic lattice of the substrate. Any displacement of the molecule's centre of mass results in a less good fit between the H atoms and the fourfold hollow sites of the Cu (110) surface.

For the rotated molecules, however, the registry between the H atoms and the fourfold hollow sites of the substrate is not as good: the molecular shape no longer fits the atomic registry of the substrate template. In this case, mechanical excitations would be expected to result in molecular displacements, because a displacement of the centre of mass does not significantly alter the registry between the H atoms and the fourfold hollow site.

The dynamics of large organic molecules on surfaces is, therefore, analogous to the dynamics of the so-called lock-and-key recognition between enzymes and the substrate on which they act. In the case of enzymes, this binding takes place only when the binding geometry of the substrate (key) makes an exact fit to the enzyme's active site (lock): any other binding geometry would decrease the fit between the enzyme and the substrate, thereby decreasing the binding energy.

This alternative binding geometry can thus be considered as a metastable state that will either decay into the correct binding geometry, provided that there are no kinetic limitations, or lead to a dissociation of the dimer, and the substrate molecule would eventually be released and drift away from the active centre. In the case of the VL molecules, when deposited at high temperature (no kinetic stabilization of metastable states) all VL molecules are oriented parallel to the close-packed direction in such a way that their shape fits the atomic lattice of the Cu (110) surface (lock). A metastable configuration, in which the molecules are rotated and therefore do not fit the Cu (110) substrate, can be forced with the STM tip and stabilized at low temperatures. The interaction between the substrate and the rotated molecules is markedly decreased by the lack of registry with the substrate template and the molecules thus move from their adsorption sites, that is, diffuse on the surface.

The results obtained by Otero et al. indicate that the mechanical stability of complex molecular species on solid surfaces depends strongly on the complimentarily between the molecular 3D shape and the atomic registry of the substrate. Molecules with a shape that fits the atomic lattice of the substrate have a very low diffusivity and therefore such molecular nanostructures are stable for longer time spans. The results are thus of great relevance to the emerging area of nanotechnology in which functional molecular nanostructures are built in a bottom-up approach.

Catalysis

We use chemical vapor deposition (CVD) for making nanoscale materials such as carbon nanotubes (CNTs) and carbon nanofibers (CNFs). In CVD we require low temperature so as to enable the use of more sensitive substrates or integration processes. We can study two types of CVD to see the importance of surface diffusion regarding this aspect.

a) In thermal CVD activation energy has to be equal to dissociation of gas phase C_2H_2 or CH_4 which produces C atom for diffusion on or in the catalyst.

b) In Plasma enhanced CVD (PECVD) the rate limiting step is carbon diffusion on catalyst surface which lowers the activation energy .Hence we can have lower temperature.

Soldering

The surface diffusion is the mechanism responsible for the transport of Sn inside the solder joint. It also causes the Sn transport rate at locations near the surface to be higher than those in the interior of solder joints. Surface diffusion also accelerates the homogenization process of the solder joints.

Mostly Sn-Pb alloy is used but due to environmental concerns arising from the toxicity of Pb, several other solders such as Sn-Ag, Sn-Cu alloys, have been considered to replace Sn-Pb solders. Proper understanding of surface diffusion parameters such as which species will dominate diffusion is necessary as it governs wetting, joint strength and reliability.

Surface Diffusion in Nanotechnology

In nanopatterning, surface diffusion is the mechanism involved. To illustrate we see the following example.

We concentrate on Co growth on strained thin Pt films. We can have a pattern of vertical pillars fabricated by self-assembly and self-organization on the surface. Here for example a thin film of Pt can be deposited on the clean flat surface of sapphire. Due to the mismatch of the lattice constants, dislocations evolve in the Pt film when it is relaxed. These dislocations repel each other so that the equilibrium configuration is given by an equally spaced dislocation array. So we have a well-defined array. During patterning process Co atoms get diffused on Pt atoms and will start nucleating. Here it was observed that diffusion increases with tensile strain showing that preferred nucleation occurs in the regions of high tensile stress. Hence in the end we get desired pattern.

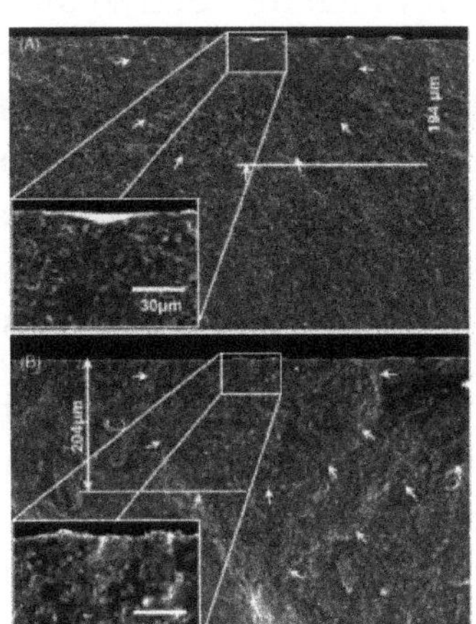

Power Metallurgy

In power metallurgy also surface diffusion plays an important role. There are three steps in power metallurgy:

- Mixing: Metal powders are first mixed with lubricants or other alloy additions to produce a homogeneous mixture of ingredients.
- Compacting: A controlled amount of mixed powder is compacted at high pressures. Compacting the loose powder produces a "green compact", which has the size and shape of the finished part and sufficient strength for in-process handling and transport to a sintering furnace.
- Sintering: Here the "green compact", is placed on a wide-mesh belt, and is slowly moved through a controlled-atmosphere furnace. Parts are heated to below the melting point of the base metal, held at the sintering temperature, and then cooled. Since we have through high-temperature treatment and sintering of powder compacts has no liquid phase. Here predominant mass transport (i.e. densification mechanism) is solid-state diffusion. We have a reduction of compact porosity. So due to surface diffusion we do not have shrinkage in volume but formation of solid bonds between adjacent particle. This results in sintered material having better mechanical properties like better strength and toughness etc.

Repairing

The mechanism of surface diffusion can be used for repairing of defects. For example as in oxidation of NiO with the matrix-Al_2O_3, the rapid diffusion along with the grain boundary repairs damages on the surface by filling and re-bonding the defects and cracks with the reaction product, $NiAl_2O_4$.

Heterogeneous Catalysis

Heterogeneous catalyst can be defined as the substances which accelerate the chemical reaction and have different phases from reactant molecules. Two different phases contains a separate boundary between them like a solid and a liquid consists of two phases.

Many catalytic processes are known in which the catalyst and the reactants are not present in the same phase—that is, state of matter. These are known as heterogeneous catalytic reactions. They include reactions between gases or liquids or both at the surface of a solid catalyst. Since the surface is the place at which the reaction occurs, it generally is prepared in ways that produce large surface areas per unit of catalyst; finely divided metals, metal gauzes, metals incorporated into supporting matrices, and metallic films have all been used in modern heterogeneous catalysis. The metals themselves are used, or they are converted to oxides, sulfides or halides.

With solid catalysts, at least one of the reactants is chemisorbed (a portmanteau term for chemically adsorbed) by the catalyst. A catalyst is chosen that releases the products formed as readily as possible; otherwise the products remain on the catalyst surface and act as poisons to the process. Chemisorption can occur over a wide temperature range, the most effective temperature for adsorption depending on the nature of the catalyst. Thus, hydrogen is chemisorbed readily by many metals even at liquid air temperatures (below −180 °C [−290 °F]). With a series of hydrogenation-dehydrogenation catalysts—e.g., zinc oxide–chromic oxide ($ZnO–Cr_2O_3$)—chemisorption of hydrogen often occurs above room temperature. Nitrogen is rapidly chemisorbed on synthetic ammonia-iron catalyst in the region above 400 °C (750 °F). It has been shown that iron films chemisorb nitrogen even at liquid air temperatures, with additional chemisorption found above room temperatures. It follows from such considerations that whereas physical adsorptions, which parallel the ease of liquefaction of the adsorbed substance, occur spontaneously, chemisorption, which involves the making and breaking of chemical bonds, often requires activation energies (energy needed to initiate reactions) as do uncatalyzed chemical processes. To be efficient catalytically, a process must involve energies of activation for all the steps involved that, at their maxima, are less than those required for the uncatalyzed reaction. This situation is illustrated graphically in the figure, with a hypothetical reaction that could occur by either an uncatalyzed or a catalyzed route.

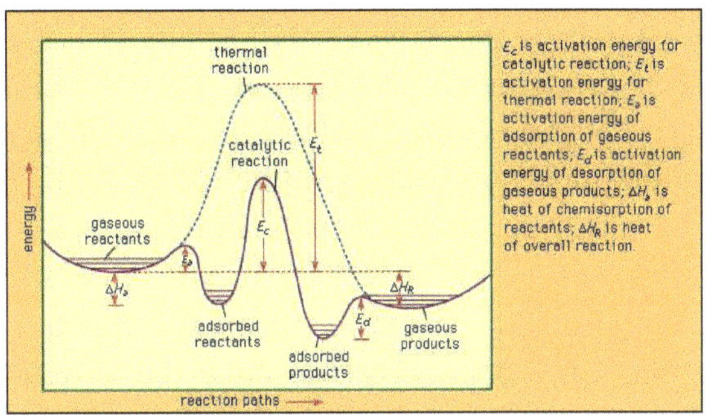

Energy profiles for catalytic and thermal (noncatalytic) reactions in the gaseous phase.

Two competing proposals have been made concerning the mechanism of catalytic reactions at surfaces, and it has not been possible to choose between them. Originally, Irving Langmuir, an American physical chemist, proposed chemisorption of both reacting species at the surface, followed by interaction between adjacent species and evaporation of the products. An alternative proposal involves interaction between an impinging molecule and species already adsorbed on the surface. Subsequent developments have suggested various modes of attachment of the adsorbed and adsorbing species.

A major advance in the science of surface catalysis was the development of a method for determining the surface area of catalysts (and other materials) by measuring the multi molecular adsorption of nitrogen at liquid nitrogen temperatures or the adsorption of other gases close to their boiling points. It then became possible to calculate a quantity, designated V_m that represents the volume of gas necessary to form a monolayer on the accessible surface; furthermore, the area of the surface can be determined from the known dimensions of the adsorbed molecules. It has also been found possible to titrate (measure quantitatively) the area of surfaces by chemisorption of gases. Since heterogeneously catalyzed reactions occur on the surface of the catalyst, the rates of such reactions are proportional to the accessible surface area of the catalyst. Active catalysts are thus usually highly porous solids with total surface areas as high as several hundred square metres per gram.

When measurements of surface areas became possible, it was seen at once that many constituents present in minor quantities in the main catalyst material—known as promoters—could act by extending the effective surface area of the catalyst. It also was shown, however, that a promoter might produce an increase in the quality of the surface for the given reaction. Acting in a reverse direction are minor constituents of the reacting system or unwanted products of the reaction, which by preferential adsorption on the reaction sites and resistance to removal give rise to poisons for the process. Poisoning of a catalyst may also result from the poison adversely modifying the electronic properties of the catalyst.

Much can be learned about mechanisms of surface processes by studying the behaviour of isotopic species of the reactants and products on the catalyst. An example of such use concerns the technically important synthesis of ammonia from its elements, the well-known Haber-Bosch process on promoted-iron catalysts. The synthesis of ammonia involves three types of bonds—hydrogen-hydrogen, nitrogen-hydrogen, and nitrogen-nitrogen—all of which can be studied using isotopes of hydrogen and nitrogen. The first of these can be examined on the reduced-iron catalyst by following the progress of the reaction $H_2 + D_2 \rightarrow 2HD$ (in which D equals the deuterium atom, an isotope of hydrogen) on the surface. The reaction is found to occur rapidly even at liquid air temperatures. Nitrogen-hydrogen bond activation can be studied by following the reaction $NH_3 + ND_3 \rightarrow NH_2D + NHD_2$. This proceeds steadily at room temperatures. The reaction involving only N–N bonds, however, studied by following the process $^{14}N_2 + {}^{15}N_2 \rightarrow 2{}^{14}N{-}{}^{15}N$ (in which ^{14}N and ^{15}N are stable isotopes of nitrogen), is shown to proceed only at the higher temperatures of ammonia synthesis, around 400 °C (750 °F). From these data one concludes that the activation of the nitrogen molecules is the slow step (the process that limits the overall reaction) in ammonia synthesis. This conclusion is confirmed by measurement of rate of adsorption of nitrogen on the iron catalysts. Other, similar isotopic studies have yielded valuable information on the reactions of hydrocarbons, using deuterium and carbon-14 as the isotopic tracers.

Mechanism of Heterogeneous Catalysis

According to Surface adsoprtion theory heterogeneous catalysis has five stages:

- Stage 1: Diffusion of Reactant(s) to the Surface- The rate at which reactants will diffuse to the surface will be influenced by their bulk concentration and by the thickness of the boundary layer.
- Stage 2: Adsorption of reactants- Bonds are formed as the reactant(s) are adsorbed onto the surface of the catalyst. The ability for an atom or molecule to stick to the surface is known, brilliantly, as the Sticking Co-efficient. This is just the ratio or percentage of molecules that end up sticking on the surface.
- Stage 3: Reaction- Bonds form between the atoms and molecules on the surface.
- Stage 4: Desorption of products- Bonds are broken as the product(s) desorb from the surface.
- Stage 5: Diffusion of Product(s) away from the Surface- The products are then desorbed from the surface of the catalyst.

Example of Contact Process

The contact process is the current method of producing sulfuric acid in the high concentrations needed for industrial processes. Platinum used to be the catalyst for this reaction; however, as it is susceptible to reacting with arsenic impurities in the sulfur feedstock, vanadium(V) oxide (V_2O_5) is now preferred.

In the contact process vanadium [V] oxide (V_2O_2) is solid whereas the reactants SO_2 and O_2 are gaseous:

$$2SO_2(g)+O_2(g)\overset{V_2O_5(s)}{\rightleftharpoons}2SO_3(g)$$

Detailed method:

$$2V_2O_5(s)+2SO_2(g)\rightleftharpoons 2SO_3(g)+2V_2O_4(s)$$

$$2V_2O_4(s)+O_2(g)\rightleftharpoons 2V_2O_5(s)$$

Therefore,

$$2SO_2(g)+O_2(g)\rightleftharpoons 2SO_3(g).$$

Example of Haber–Bosch Reaction

The original Haber–Bosch reaction chambers used osmium as the catalyst, less expensive iron-based catalyst, which is still used today.

$$N_2(g)+3H_2(g)\overset{Fe(s)}{\rightleftharpoons}2NH_3(g)$$

The reaction mechanism, involving the heterogeneous catalyst, is believed to involve the following steps:

1. N_2 (g) → N_2 (adsorbed)
2. N_2 (adsorbed) → 2 N (adsorbed)
3. H_2 (g) → H_2 (adsorbed)
4. H_2 (adsorbed) → 2 H (adsorbed)
5. N (adsorbed) + 3 H (adsorbed)→ NH_3 (adsorbed)
6. NH_3 (adsorbed) → NH_3 (g).

Heterogeneous Gold Catalysis

Heterogeneous gold catalysis refers to the catalysis of chemical reactions by gold nanoparticles, typically supported on metal oxide substrates. Despite the well known inertness of bulk gold, decreasing the diameter of supported gold clusters to ca. 2 to 5 nm has been shown to result in surprisingly high catalytic activities towards low-temperature carbon monoxide (CO) oxidation as well as several other industrially relevant reactions such as H_2 activation, water gas shift and hydrogenation of unsaturated hydrocarbons. The high activity of supported gold clusters has been proposed to arise from a combination of structural changes, quantum-size effects and support effects that preferentially tune the electronic structure of gold such that optimal binding of adsorbates during the catalytic cycle is enabled. The selectivity and activity of gold nanoparticles can be finely tuned by varying the choice of support material, with e.g. titania (TiO_2), hematite (α-Fe_2O_3), cobalt(II/III) oxide (Co_3O_4) and nickel(II) oxide (NiO) serving as the most effective support materials for facilitating the catalysis of CO combustion. Besides enabling an optimal dispersion of the nanoclusters, the support materials have been suggested to promote catalysis by altering the size, shape, strain and charge state of the cluster. A precise shape control of the deposited gold clusters has been shown to be important for optimizing the catalytic activity, with hemispherical, few atomic layers thick nanoparticles generally exhibiting the most desirable catalytic properties due to maximized number of high-energy edge and corner sites.

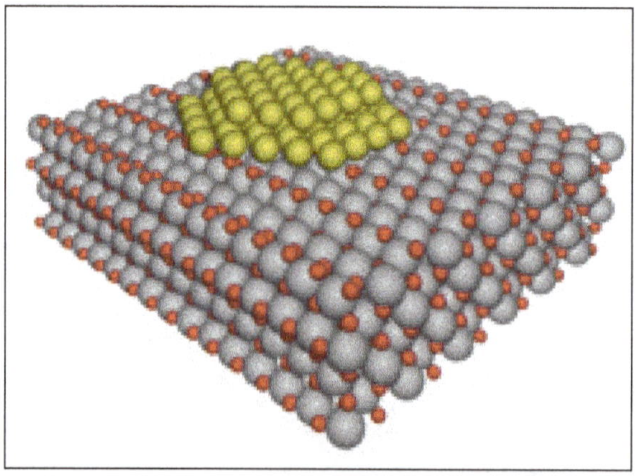

Model structure of a 75-atom two-layer gold cluster with a diameter of roughly 2 nm deposited on a rutile TiO_2(110) surface. Yellow, gray and red spheres depict gold, titanium and oxygen atoms, respectively. The structure has not been optimized computationally, and thus serves only as a conceptual illustration.

Bulk metallic gold is known to be inert, exhibiting a surface reactivity at room temperature only towards a few substances such as formic acid and sulphur-containing compounds, e.g. H2S and thiols. Within heterogeneous catalysis, reactants adsorb onto the surface of the catalyst thus forming activated intermediates. However, if the adsorption is weak such as in the case of bulk gold, a sufficient perturbation of the reactant electronic structure does not occur and catalysis is hindered (Sabatier's principle). When gold is deposited as nanosized clusters of less than 5 nm onto metal oxide supports, a markedly increased interaction with adsorbates is observed, thereby resulting in surprising catalytic activities. Evidently, nano-scaling and dispersing gold on metal oxide substrates makes gold less noble by tuning its electronic structure, but the precise mechanisms underlying this phenomenon are as of yet uncertain and hence widely studied.

It is generally known that decreasing the size of metallic particles in some dimension to the nanometer scale will yield clusters with a significantly more discrete electronic band structure in comparison with the bulk material. This is an example of a quantum-size effect and has been previously correlated with an increased reactivity enabling nanoparticles to bind gas phase molecules more strongly. In the case of TiO_2-supported gold nanoparticles, Valden *et al.* observed the opening of a band gap of approximately 0.2-0.6 eV in the gold electronic structure as the thickness of the deposited particles was decreased below three atomic layers. The two-layer thick supported gold clusters were also shown to be exceptionally active for CO combustion, based on which it was concluded that quantum-size effects inducing a metal-insulator transition play a key role in enhancing the catalytic properties of gold. However, decreasing the size further to a single atomic layer and a diameter of less than 3 nm was reported to again decrease the activity. This has later been explained by a destabilization of clusters composed of very few atoms, resulting in too strong bonding of adsorbates and thus poisoning of the catalyst.

The properties of the metal d-band are central for describing the origin of catalytic activity based on electronic effects. According to the d-band model of heterogeneous catalysis, substrate-adsorbate bonds are formed as the discrete energy levels of the adsorbate molecule interacts with the metal d-band, thus forming bonding and antibonding orbitals. The strength of the formed bond depends on the position of the d-band center such that a d-band closer to the Fermi level (E_F) will result in stronger interaction. The d-band center of bulk gold is located far below E_F, which qualitatively explains the observed weak binding of adsorbates as both the bonding and antibonding orbitals formed upon adsorption will be occupied, resulting in no net bonding. However, as the size of gold clusters is decreased below 5 nm, it has been shown that the d-band center of gold shifts to energies closer to the Fermi level, such that the as formed antibonding orbital will be pushed to an energy above E_F, hence reducing its filling. In addition to a shift in the d-band center of gold clusters, the size-dependency of the d-band width as well as the $5d_{3/2}$-$d_{5/2}$ spin-orbit splitting has been studied from the viewpoint of catalytic activity. As the size of the gold clusters is decreased below 150 atoms (diameter ca. 2.5 nm), rapid drops in both values occur. This can be attributed to d-band narrowing due to the decreased number of hybridizing valence states of small clusters as well as to the increased ratio of high-energy edge atoms with low coordination to the total number of Au atoms. The effect of the decreased $5d_{3/2}0d_{5/2}$ spin-orbit splitting as well as the narrower distribution of d-band states on the catalytic properties of gold clusters cannot be understood via simple qualitative arguments as in the case of the d-band center model. Nevertheless, the observed trends provide further evidence that a significant perturbation of the Au electronic structure occurs upon nanoscaling, which is likely to play a key role in the enhancement of the catalytic properties of gold nanoparticles.

A central structural argument explaining the high activity of metal oxide supported gold clusters is based on the concept of periphery sites formed at the junction between the gold cluster and the substrate. In the case of CO oxidation, it has been hypothesized that CO adsorbs onto the edges and corners of the gold clusters, while the activation of oxygen occurs at the peripheral sites. The high activity of edge and corner sites towards adsorption can be understood by considering the high coordinative unsaturation of these atoms in comparison with terrace atoms. The low degree of coordination increases the surface energy of corner and edge sites, hence making them more active towards binding adsorbates. This is further coupled with the local shift of the d-band center of the unsaturated Au atoms towards energies closer to the Fermi level, which in accordance with the d-band model results in increased substrate-adsorbate interaction and lowering of the adsorption-dissociation energy barriers. Lopez *et al.* calculated the adsorption energy of CO and O_2 on the Auterrace on which the Au-atoms have a coordination number of 9 as well as on an Au_{10} cluster where the most reactive sites have a coordination of 4. They observed that the bond strengths are in general increased by as much as 1 eV, indicating a significant activation towards CO oxidation if one assumes that the activation barriers of surface reactions scale linearly with the adsorption energies (Brønsted-Evans-Polanyi principle). The observation that hemispherical two-layer gold clusters with a diameter of a few nanometers are most active for CO oxidation is well in line with the assumption that edge and corner atoms serve as the active sites, since for clusters of this shape and size the ratio of edge atoms to the total number of atoms is indeed maximized.

The preferential activation of O_2 at the perimeter sites is an example of a support effect that promotes the catalytic activity of gold nanoparticles. Besides enabling a proper dispersion of the deposited particles and hence a high surface-to-volume ratio, the metal oxide support also directly perturbs the electronic structure of the deposited gold clusters via various mechanisms, including strain-inducing and charge transfer. For gold deposited on magnesia (MgO), a charge transfer from singly charged oxygen vacancies (F-centers) at the MgO surface to the Au cluster has been observed. This charge transfer induces a local perturbation in the electronic structure of the gold clusters at the perimeter sites, enabling the formation of resonance states as the antibonding $2\pi^*$ orbital of oxygen interacts with the metal d-band. As the antibonding orbital is occupied, the O-O bond is significantly weakened and stretched, i.e. activated. The activation of O_2 at the perimeter sites is also observed for defect-free surfaces and neutral gold clusters, but to a significantly smaller extent. The activity enhancing effect of charge transfer from the substrate to gold has also been reported by Chen and Goodman in the case of a gold bilayer supported on ultrathin TiO_2 on Mo. In addition to charge transfer between the substrate and the gold nanoparticles, the support material has been observed to increase the catalytic activity of gold by inducing strain as a consequence of lattice mismatch. The induced strains especially affect the Au atoms close to the substrate-cluster interface, resulting in a shift of the local d-band center towards energies closer to the Fermi level. This corroborates the periphery hypothesis and the creation of catalytically active bifunctional sites at the cluster-support interface. Furthermore, the support-cluster interaction directly influences the size and shape of the deposited gold nanoparticles. In the case of weak interaction, less active 3D clusters are formed, whereas if the interaction is stronger more active 2D few-layer structures are formed. This illustrates the ability to fine-tune the catalytic activity of gold clusters via varying the support material as well as the underlying metal upon which the substrate has been grown.

Finally, it has been observed that the catalytic activity of supported gold clusters towards CO oxidation is further enhanced by the presence of water. Invoking the periphery hypothesis, water promotes the activation of O_2 by co-adsorption onto the perimeter sites where it reacts with O_2 to form adsorbed hydroxyl (OH*) and hydroperoxo (OOH*) species. The reaction of these intermediates with adsorbed CO is very rapid, and results in the efficient formation of CO_2 with concomitant recovery of the water molecule.

Applications

CO Oxidation

Gold can be a very active catalyst in oxidation of carbon monoxide (CO), i.e. the reaction of CO with molecular oxygen to produce carbon dioxide (CO_2). Supported gold clusters, thin films and nanoparticles are one to two orders of magnitude more active than atomically dispersed gold cations or unsupported metallic gold.

Gold cations can be dispersed atomically on basic metal oxide supports such as MgO and La_2O_3. Monovalent and trivalent gold cations have been identified, the latter being more active but less stable than the former. The turnover frequency (TOF) of CO oxidation on these cationic gold catalysts is in the order of magnitude of 0.01 s^{-1}, exhibiting the very high activation energy of 138 kJ/mol.

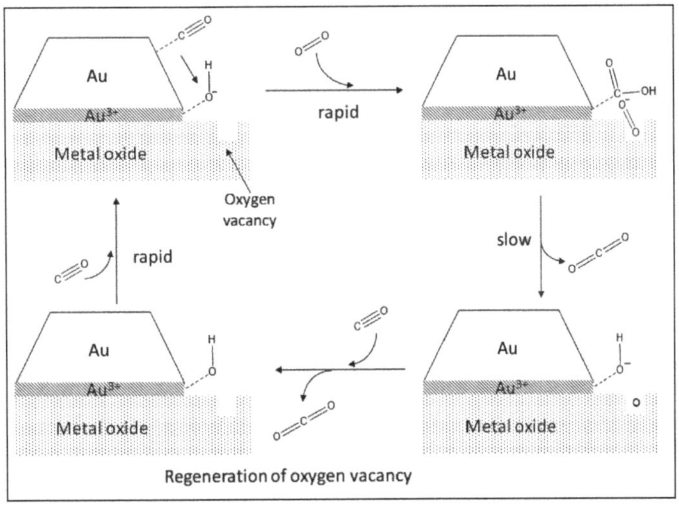

Possible mechanism of CO oxidation on an Au catalyst supported on a reducible metal oxide at room temperature. The reaction is aided by surface hydoxyl species.

Supported gold nanoclusters with a diameter < 2 nm are active to CO oxidation with TOF in the order of magnitude of 0.1 s^{-1}. It has been observed that clusters with 8 to 100 atoms are catalytically active. The reason is that, on one hand, eight atoms are the minimum necessary to form a stable, discrete energy band structure, and on the other hand, d-band splitting decreases in clusters with more than 100 atoms, resembling the bulk electronic structure. The support has a substantial effect on the electronic structure of gold clusters. Metal hydroxide supports such as $Be(OH)_2$, $Mg(OH)_2$, and $La(OH)_3$, with gold clusters of < 1.5 nm in diameter constitute highly active catalysts for CO oxidation at 200 K (-73 °C). By means of techniques such as HR-TEM and EXAFS, it has been proven that the activity of these catalysts is due exclusively to clusters with 13 atoms arranged in an icosahedron structure. Furthermore, the metal loading should exceed 10 wt% for the catalysts to be active.

Gold nanoparticles in the size range of 2 to 5 nm catalyze CO oxidation with a TOF of about 1 s^{-1} at temperatures below 273 K (0 °C). The catalytic activity of nanoparticles is brought about in the absence of moisture when the support is semiconductive or reducible, e.g. TiO_2, MnO_2, Fe_2O_3, ZnO, ZrO_2, or CeO_2. However, when the support is insulating or non-reducible, e.g. Al_2O_3 and SiO_2, a moisture level > 5000 ppm is required for activity at room temperature. In the case of powder catalysts prepared by wet methods, the surface OH^- groups on the support provide sufficient aid as co-catalysts, so that no additional moisture is necessary. At temperatures above 333 K (60 °C), no water is needed at all.

The apparent activation energy of CO oxidation on supported gold powder catalysts prepared by wet methods is 2-3 kJ/mol above 333 K (60 °C) and 26-34 kJ/mol below 333 K. These energies are low, compared to the values displayed by other noble metal catalysts (80-120 kJ/mol). The change in activation energy at 333 K can be ascribed to a change in reaction mechanism. This explanation has been supported experimentally. At 400 K (127 °C), the reaction rate per surface Au atom is not dependent on particle diameter, but the reaction rate per perimeter Au atom is directly proportional to particle diameter. This suggests that the mechanism above 333 K takes place on the gold surfaces. By contrast, at 300 K (27 °C), the reaction rate per surface Au atom is inversely proportional to particle diameter, while the rate per perimeter interface does not depend on particle size. Hence, CO oxidation occurs on the perimeter sites at room temperature. Reaction mechanism has been revealed by studying the dependency of the reaction rate on the partial pressures of the reactive species. Both at 300 K and 400 K, there is a first order rate dependency on CO partial pressure up to 4 Torr (533 Pa), above which the reaction is zero order. With respect to O_2, the reaction is zero order above 10 Torr (54.7 kPa) at both 300 and 400 K. The order with respect to O_2 at lower partial pressures is 1 at 300 K and 0.5 at 400 K. The shift towards zero order indicates that the catalyst's active sites are saturated with the species in question. Hence, a Langmuir-Hinshelwood mechanism has been proposed, in which CO adsorbed on gold surfaces reacts with O adsorbed at the edge sites of the gold nanoparticles.

The need to use oxide supports, and more specifically reducible supports, is due to their ability to activate dioxygen. Gold nanoparticles supported on inert materials such as carbon or polymers have been proven inactive in CO oxidation. The aforementioned dependency of some catalysts on water or moisture also relates to oxygen activation. The ability of certain reducible oxides, such as MnO_2, Co_3O_4, and NiO to activate oxygen in dry conditions (< 0.1 ppm H_2O) can be ascribed to the formation of oxygen defects during pretreatment.

Water Gas Shift

Water gas shift is the most widespread industrial process for the production of dihydrogen, H_2. It involves the reaction of carbon monoxide and water (syngas) to form hydrogen and carbon dioxide as a byproduct. In many catalytic reaction schemes, one of the elementary reactions is the oxidation of CO with an adsorbed oxygen species. Gold catalysts have been proposed as an alternative for water gas shift at low temperatures, viz. < 523 K (250 °C). This technology is essential to the development of solid oxide fuel cells. Hematite has been found to be an appropriate catalyst support for this purpose. Furthermore, a bimetallic Au-Ru/Fe_2O_3 catalyst has been proven highly active and stable for low-temperature water gas shift. Titania and ceria have also been used as supports for effective catalysts. Unfortunately, Au/CeO_2 is prone to deactivation caused by surface-bound carbonate or formate species.

Although gold catalysts are active at room temperature to CO oxidation, the high amounts of water involved in water gas shift require higher temperatures. At such temperatures, gold is fully reduced to its metallic form. However, the activity of e.g. Au/CeO_2 has been enhanced by CN– treatment, whereby metallic gold is leached, leaving behind highly active cations. According to DFT calculations, the presence of such Au cations on the catalyst is allowed by empty, localized nonbonding f states in CeO_2. On the other hand, STEM studies of Au/CeO_2 have revealed nanoparticles of 3 nm in diameter. Water gas shift has been proposed to occur ar the interface of Au nanoparticles and the reduced CeO_2 support.

Selective Oxidations

In petrochemistry, high-value intermediates such as carboxylic acids, aldehydes, ketones, epoxides, and alcohols are obtained by partial oxidation of alkanes and alkenes with dioxygen. These intermediates are essential to the production of consumer goods. Partial oxidation presents two challenges. The first is that the most favored reaction between oxygen and hydrocarbons is combustion. The second challenge is the considerable difficulty to activate dioxygen, viz. the splitting of the molecule into its constituent atoms, which has an energy barrier of 498 kJ/mol. The usual strategy to activate oxygen in a controlled manner is to use molecular hydrogen or carbon monoxide as sacrificial reductants in presence of a heterogeneous catalyst, such that the activation barrier is lowered to < 10 kJ/mol and hence milder reaction conditions are required.

One of the most challenging selective oxidations, which have been achieved with supported gold catalysts, is the epoxidation of propylene. Although the epoxidation of ethylene is routinely achieved in the industry with selectivities as high as 90% on Ag catalysts, most catalysts provided < 10% selectivity for propylene epoxidation. Using a gold catalyst supported on titanium silicate-1 (TS-1) molecular sieve, yields of 350 g/h per gram of gold were obtained at 473 K (200 °C). The reaction took place in the gas phase. Furthermore, using mesoporous titanosilicate supports (Ti-MCM-41 and Ti-MCM-48), gold catalysts provided > 90% selectivity at ~ 7% propylene conversion, 40% H_2 efficiency, and 433 K (160 °C). The active species in these catalysts were identified to be hemispherical gold nano-crystals of less than 2 nm in diameter in intimate contact with the support.

Alkene epoxidation has been demonstrated in absence of H_2 reuductant in the liquid phase. For example, using 1% Au/graphite, ~80% selectivities of cis-cyclooctene to cyclooctene oxide (analogous to cyclohexene oxide) were obtained at 7-8% conversion, 353 K (80 °C), and 3 MPa O_2 in absence of hydrogen or solvent. Other liquid-phase selective oxidations have been achieved with saturated hydrocarbons. For instance, cyclohexane has been converted to cyclohexanone and cyclohexanol with a combined selectivity of ~100% on gold catalysts. Product selectivities can be tuned in liquid phase reactions by the presence or absence of solvent and by the nature of the latter, viz. water, polar, or nonpolar. With gold catalysts, the catalyst's support has less influence on reactions in the liquid phase than on reactions in the gas phase.

Selective Hydrogenations

Typical hydrogenation catalysts are based on metals from the 8, 9, and 10 groups, such as Ni, Ru, Pd, and Pt. By comparison, gold has a poor catalytic activity for hydrogenation. This low activity is caused by the difficulty of dihydrogen activation on gold. While hydrogen dissociates on Pd and Pt without an energy barrier, dissociation on Au has an energy barrier of ~1.3 eV, according to DFT calculations. These calculations agree with experimental studies, in which no hydrogen dissociation was observed on gold or terraces, nor on steps. No dissociation was observed on these surfaces either at room temperature or at 473 K (200 °C). However, the rate of hydrogen activation increases as the Au nanoparticles decrease in diamerer at 423 K (150 °C). Furthermore, hydrogen dissociation rate per periphery Au atom remains constant when the particle size is changed, suggesting that activation occurs at the periphery of Au nanoparticles. Notwithstanding its poor activity, nano-sized gold immovilized in various supports has been found to provide a good selectivity in hydrogenation reactions.

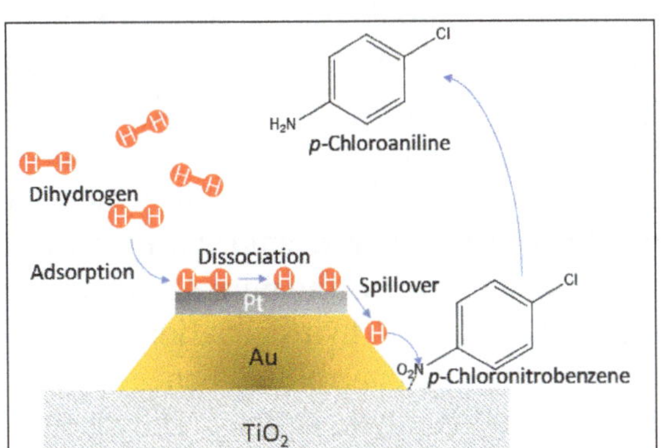

Schematic of the reaction mechamism of the selective hydrogenation of *p*-chloronitrobenzene on a bimetallic Pt. -Au/TiO_2 catalyst.

One of the early studies of hydrogenation on supported, highly dispersed gold was performed with 1-butene and cyclohexene in the gas phase at 383 K (110 °C). The reaction rate was found to be first order with trespect to alkene pressure and second order with respect to chemisorbed hydrogen. In later works, it was shown that gold-catalyzed hydrogenation can be highly sensitive to Au loading (hence to particle size) and to the nature of the support. For example, 1-pentene hydrogenation occurred optimally on 0.04 wt% Au/SiO_2, but not at all on $Au/\gamma\text{-}Al_2O_3$. By contrast, the hydrogenation of 1,3-butadiene to 1-butene was shown to be relatively insensitive to Au particle size in a study with a series of Au/Al_2O_3 catalysts prepared by different methods. With all the tested catalysts, conversion was ~100% and selectivity, < 60%. Concerning reaction mechanisms, in a study of propylene hydrogenation on Au/SiO_2, reaction rates were determined using D_2 and H_2. Because the reaction with deuterium was substantially slower, it was suggested that the rate-determining step in alkene hydrogenation was the cleavage of the H-H bond. Lastly, ethylene hydrogenation was studied on Au/MgO at atmospheric pressure and 353 K (80 °C) with EXAFS, XANES and IR spectroscopy, suggesting that the active species might be Au^{+3} and the reaction intermediate, an ethylgold species.

Gold catalysts are especially selective in the hydrogenation of α,β-insaturated aldehydes, i.e. aldehydes containing a C=C double bond on the carbon adjacent to the carbonyl. Gold catalysts are able to hydrogenate only the carbonyl group, so that the aldehyde is transformed to the corresponding alcohol, while leaving the C=C double bond untouched. In the hydrogenation of crotonaldehyde to crotyl alcohol, 80% selectivity was attained at 5-10% conversion and 523 K (250 °C) on Au/ZrO_2 and Au/ZnO. The selectivity increased along with Au particle size in the range of ~2 to ~5 nm. Other instances of this reaction include acrolein, citral, benzal acetone, and pent-3-en-2-one. The activity and selectivity of gold catalysts for this reaction has been linked to the morphology of the nanoparticles, which in turn is influenced by the support. For example, round partilces tend to form on TiO_2, while ZnO promotes particles with clear facets, as observed by TEM. Because the round morphology provides a higher relative amount of low-coordinated metal surface sites, the higher activity observerd with Au/TiO_2 compared to Au/ZnO is explained. Finally, a bimetallic Au-In/ZnO catalyst has been observed to improve the selectivity towards the hydrogenation of the carbonyl in acrolein. It was observed in HRTEM images that indium thin films decorate some of the facets of the gold nanoparticle. The promoting effect on selectivity might result from the fact that only the Au sites that promote side-reactions are decorated by In.

A strategy that in many reactions has succeeded at improving gold's catalytic activity without impairing its selectivity is to synthesize bimetallic Pd-Au or Pt-Au catalysts. For the hydrogenation of 1,3-butadiene to butenes, model surfaces of Au, Pd-Au, Pd-Au, and Pd were studied with LEED, AES, and LEIS. A selectivity of ~100% was achieved on $Pd_{70}Au_{30}$ and it was suggested that Au might promote the desorption of the product during the reaction. A second instance is the hydrogenation of *p*-chloronitrobenzene to *p*-chloroaniline, in which selectivity suffers with typical hydrogenation catalysts due to the parallel hydrodechlorination to aniline. However, $Pd\text{-}Au/Al_2O_3$ (Au/Pd ≥20) has been proven thrice as active as the pure Au catalyst, while being ~100% selective to *p*-chloroaniline. In a mechanistic study of hydrogenation of nitrobenzenes with $Pt\text{-}Au/TiO_2$, the dissociation of H_2 was identified as rate-controlling, hence the incorporation of Pt, an efficient hydrogenation metal, highly improved catalytic activity. Dihydrogen dissociated on Pt and the nitroaromatic compound was activated on the $Au\text{-}TiO_2$ interface. Finally, hydrogenation was enabled by the spillover of activated H surface species from Pt to the Au surface.

Colloids

A colloid is one of the three primary types of mixtures, with the other two being a solution and suspension. A colloid is a solution that has particles ranging between 1 and 1000 nanometers in diameter, yet are still able to remain evenly distributed throughout the solution. These are also known as colloidal dispersions because the substances remain dispersed and do not settle to the bottom of the container. In colloids, one substance is evenly dispersed in another. The substance being dispersed is referred to as being in the dispersed phase, while the substance in which it is dispersed is in the continuous phase.

Properties of Colloids

To be classified as a colloid, the substance in the dispersed phase must be larger than the size of a molecule but smaller than what can be seen with the naked eye. This can be more precisely quantified as one or more of the substance's dimensions must be between 1 and 1000 nanometers. If the dimensions are smaller than this the substance is considered a solution and if they are larger than the substance is a suspension.

Classifying Colloids

A common method of classifying colloids is based on the phase of the dispersed substance and what phase it is dispersed in. The types of colloids include sol, emulsion, foam, and aerosol.

1. Sol is a colloidal suspension with solid particles in a liquid.
2. Emulsion is between two liquids.
3. Foam is formed when many gas particles are trapped in a liquid or solid.
4. Aerosol contains small particles of liquid or solid dispersed in a gas.

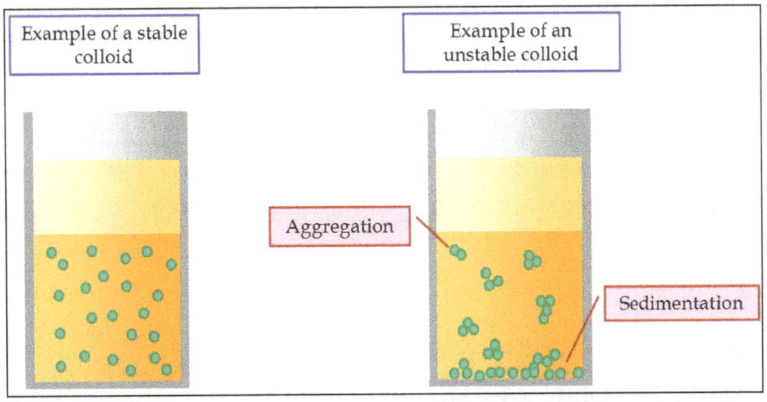

Figure: Examples of a stable and of an unstable colloidal dispersion.

When the dispersion medium is water, the collodial system is often referred to as a hydrocolloid. The particles in the dispersed phase can take place in different phases depending on how much water is available. For example, Jello powder mixed in with water creates a hydrocolloid. A common use for hydrocolloids is in the creation of medical dressings.

Table: Examples of Colloids.

Dispersion Medium	Dispersed Phase	Type of Colloid	Example
Solid	Solid	Solid sol	Ruby glass
Solid	Liquid	Solid emulsion/gel	Pearl, cheese
Solid	Gas	Solid foam	Lava, pumice
Liquid	Solid	Sol	Paints, cell fluids
Liquid	Liquid	Emulsion	Milk, oil in water
Liquid	Gas	Foam	Soap suds, whipped cream
Gas	Solid	Aerosol	Smoke
Gas	Liquid	Aerosol	Fog, mist

An easy way of determining whether a mixture is colloidal or not is through use of the Tyndall Effect. When light is shined through a true solution, the light passes cleanly through the solution, however when light is passed through a colloidal solution, the substance in the dispersed phases scatters the light in all directions, making it readily seen. An example of this is shining a flashlight into fog. The beam of light can be easily seen because the fog is a colloid.

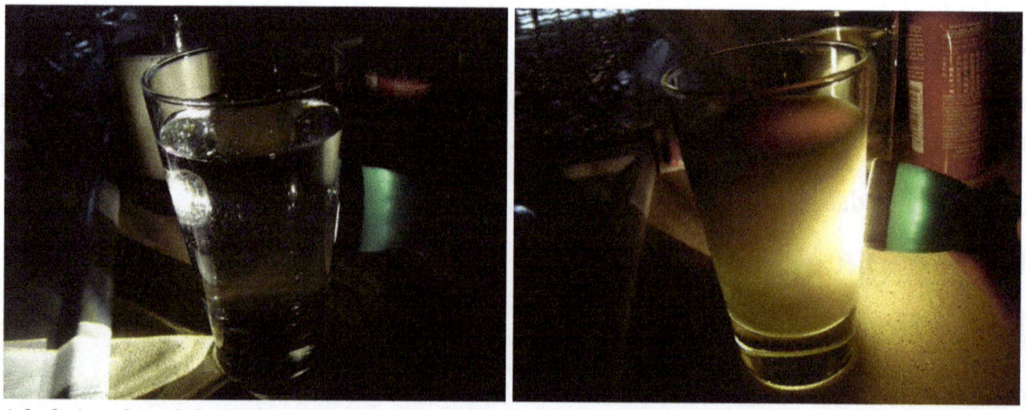

Figure: Light being shined through water and milk. The light is not reflected when passing through the water because it is not a colloid. It is however reflected in all directions when it passes through the milk, which is colloidal.

Another method of determining whether a mixture is a colloid is by passing it through a semipermeable membrane. The larger dispersed particles in a colloid would be unable to pass through the membrane, while the surrounding liquid molecules can. Dialysis takes advantage of the fact that colloids cannot diffuse through semipermeable membranes to filter them out of a medium.

Methods of Preparation

There are two principal ways of preparation of colloids:

- Dispersion of large particles or droplets to the colloidal dimensions by milling, spraying, or application of shear (e.g., shaking, mixing, or high shear mixing).
- Condensation of small dissolved molecules into larger colloidal particles by precipitation, condensation, or redox reactions. Such processes are used in the preparation of colloidal silica or gold.

Stabilization of Colloids

The colloid is said to be stable when particles remain suspended in the solution without settling down i.e. the dispersed phase. Stability is hindered by aggregation and sedimentation phenomena, which are driven by the colloid's tendency to reduce surface energy. In order to stabilize the colloidal system, we need to reduce the interfacial tension between the colloidal particles.

Aggregation is due to the sum of the interaction forces between particles. If attractive forces (such as van der Waals forces) prevail over the repulsive ones (such as the electrostatic ones) particles aggregate in clusters. Electrostatic stabilization and steric stabilization are the two main mechanisms for stabilization against aggregation.

- Electrostatic stabilization is based on the mutual repulsion of like electrical charges. In general, different phases have different charge affinities, so that an electrical double layer forms at any interface. Small particle sizes lead to enormous surface areas, and this effect is greatly amplified in colloids. In a stable colloid, the mass of a dispersed phase is so low that its buoyancy or kinetic energy is too weak to overcome the electrostatic repulsion between charged layers of the dispersing phase.
- Steric stabilization consists in covering the particles in polymers which prevents the particle to get close in the range of attractive forces.

Applications of Colloids

Colloids have varied applications. Some of them include:

- Medicines: Medicines in colloidal form are easily absorbed by the body tissues and hence are more effective.
- Cleansing action of soap: Soap solution is colloidal in nature. It removes the dirt particles either by adsorption or by emulsifying the greasy matter sticking to the cloth.
- Purification of water: The precipitation of colloidal impurities present in water can be done by adding certain electrolytes like alum etc. The negatively charged colloidal particles of impurities get neutralized by the Al^{3+} ions and settle down and pure water can be decanted off.
- Rubber industry: Latex is a colloidal solution of negatively charged rubber particles. From latex, rubber can be obtained by coagulation. Rubber plated articles are prepared by depositing negatively charged rubber particles over the article to be rubber plated by making that article an anode in a rubber plating bath.

Tyndall Effect

The Tyndall effect is the phenomenon in which the particles in a colloid scatter the beams of light that are directed at them. This effect is exhibited by all colloidal solutions and some very fine suspensions. Therefore, it can be used to verify if a given solution is a colloid. The intensity of scattered light depends on the density of the colloidal particles as well as the frequency of the incident light.

When a beam of light passes through a colloid, the colloidal particles present in the solution do not allow the beam to completely pass through. The light collides with the colloidal particles and is scattered (it deviates from its normal trajectory, which is a straight line).

This scattering makes the path of the light beam visible, as illustrated below:

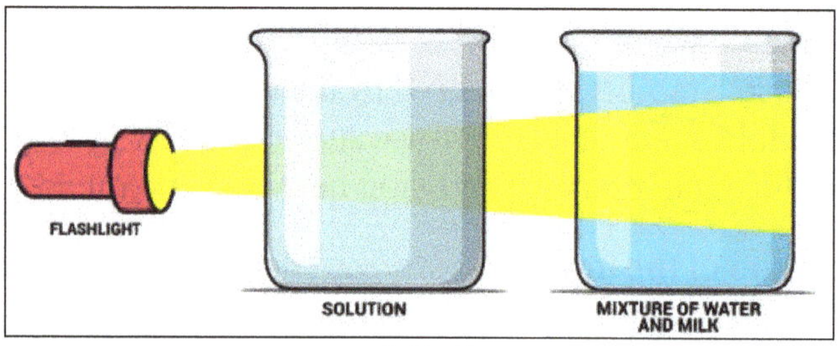

Tyndall effect

Generally, blue light is scattered to a greater extent when compared to red light. This is because the wavelength of blue light is smaller than that of red light. This is the reason why the smoke released by motorcycles sometimes appears blue.

The Tyndall effect was first discovered by (and is named after) the Irish physicist John Tyndall. The diameters of the particles that cause the Tyndall effect can range from 40 to 900 nanometers (1 nanometer = 10^{-9} meter). In comparison, the wavelength of the visible light spectrum ranges from 400 to 750 nanometers.

Examples of the Tyndall Effect:

- Milk is a colloid that contains globules of fat and protein. When a beam of light is directed at a glass of milk, the light is scattered. This is a great example of the Tyndall effect.
- When a torch is switched on in a foggy environment, the path of the light becomes visible. In this scenario, the water droplets in the fog are responsible for the light scattering.
- Opalescent glass has a bluish appearance when viewed from the side. However, orange-colored light emerges when light is shined through the glass.

How is the Tyndall Effect Responsible for Blue Eye Color?

The primary difference between blue, brown, and black colored irises is the amount of melanin in one of its layers. The layer in a blue iris has relatively lower amounts of melanin in it when compared to a black iris, making it translucent. When light is incident on this translucent layer, it is scattered due to the Tyndall effect.

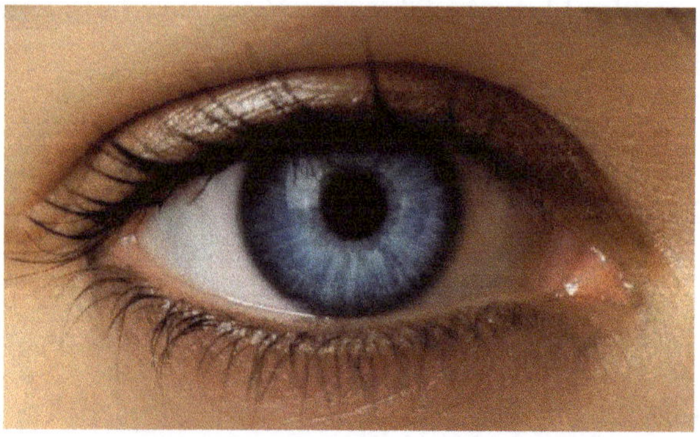

Since blue light has a shorter wavelength when compared to red light, it is scattered to a greater extent. Another layer deeper in the iris absorbs the unscattered light. Since the majority of the scattered light is blue, these irises gain their characteristic blue color.

Several phenomena involve the scattering of light. Rayleigh scattering and Mie scattering are examples of such phenomena. Clear sky is blue in color due to the scattering of light by air particles, which is an example of Rayleigh scattering. However, when the sky is cloudy, the relatively large cloud droplets are responsible for the scattering of light, which is an example of Mie scattering.

References

- Adsorption, surface-chemistry, chemistry: emedicalprep.com, Retrieved 11 April, 2019
- Langmuir-adsorption-isotherm: chemistrylearning.com, Retrieved 30 July, 2019
- Heterogeneous-catalysis, catalysis, science: britannica.com, Retrieved 1 August, 2019
- Brunauer, Stephen; Emmett, P. H.; Teller, Edward (1938). "Adsorption of Gases in Multimolecular Layers". Journal of the American Chemical Society. 60 (2): 309–319. Bibcode:1938JAChS..60..309B. doi:10.1021/ja01269a023. ISSN 0002-7863
- Heterogeneous-catalysis, Catalysis, Inorganic-Chemistry: libretexts.org, Retrieved 7 February, 2019
- Colloid, Solutions-and-Mixtures, Physical-Properties-of-Matter, Physical-and-Theoretical-Chemistry: libretexts.org, Retrieved 25 May, 2019
- Van Santen, Rutger Anthony; Neurock, Matthew (2006). Molecular heterogeneous catalysis. A conceptual and computational approach. Weinheim, Germany: Wiley-VCH. pp. 53–60. ISBN 978-3-527-29662-0
- Colloids, surface-chemistry, chemistry: toppr.com, Retrieved 5 January, 2019
- Tyndall-effect-dispersion-of-light, chemistry: byjus.com, Retrieved 15 June, 2019

4

Quantum Chemistry

The sub-field of chemistry which deals with the application of quantum mechanics in experiments and physical models of chemical systems is known as quantum chemistry. Some of concepts which are studied within it are electronic structure and chemical dynamics. This chapter discusses in detail these theories and concepts related to quantum chemistry.

Quantum chemistry is the application of quantum mechanical principles and equations to the study of molecules. In order to understand matter at its most fundamental level, we must use quantum mechanical models and methods. There are two aspects of quantum mechanics that make it different from previous models of matter. The first is the concept of wave-particle duality; that is, the notion that we need to think of very small objects (such as electrons) as having characteristics of both particles and waves. Second, quantum mechanical models correctly predict that the energy of atoms and molecules is always quantized, meaning that they may have only specific amounts of energy. Quantum chemical theories allow us to explain the structure of the periodic table, and quantum chemical calculations allow us to accurately predict the structures of molecules and the spectroscopic behavior of atoms and molecules.

Wave-Particle Duality for Light

Quantum mechanical ideas began with studies of the physics of light. By the late nineteenth century, virtually all scientists believed that light behaved as a wave. Although some earlier scientists, such as Isaac Newton in the seventeenth century, had thought of light as consisting of particles, the early nineteenth-century experiments of Thomas Young and Augustin Fresnel demonstrated that light has wavelike properties. In these experiments, light was passed through a pair of slits in a screen, and produced alternating light and dark regions (interference patterns) on a second screen. This phenomenon, known as diffraction, cannot be explained using a particle model for light. In the late nineteenth century, James Clerk Maxwell derived a set of equations based on the wave model for light, which beautifully explained most experimental results.

Despite this apparent certainty that light was a wave, Max Planck and Albert Einstein, at the beginning of the twentieth century, showed that some experiments required the use of a particle model for light, rather than a wave model. Since both models were necessary for an accurate description of all of the properties of light, scientists today use mathematical equations appropriate to both waves and particles in describing the properties of light.

Waves and particles are fundamentally different: a particle exists at a particular point in space, whereas a wave continues on for (sometimes) a great distance. It defies intuition to think that both of these models might describe the same thing. Nevertheless, an accurate description of light requires the use of both wave and particle ideas.

The Wave Nature of Matter

The success of wave-particle duality in describing the properties of light paved the way for using that same idea in describing matter. Experiments in the early twentieth century showed that the energy in atoms is quantized; that is, a given atom can have only specific amounts of energy. For hydrogen, the simplest of the atoms, an accurate formula for the possible energies had been experimentally determined but was unexplainable using any particle model for the atom. The best picture that the particle model could give, consistent with experiments on atoms, put the electron in a sort of "orbit" around the nucleus. Unfortunately, the particle model predicts that the electron should collide with the nucleus, releasing energy in the process. Obviously there was a need for a different model for the electron.

In 1924 Louis de Broglie presented a theory for the hydrogen atom that modeled the electron as a wave. Calculations made for this model give the quantization of energy that is experimentally observed in this atom. De Broglie also postulated a general formula for obtaining the wavelength of a moving object. His formula, which is analogous to that used for light, states that the wavelength of a moving object is inversely proportional to its momentum (mass times velocity). When one uses de Broglie's formula to determine wavelengths of macroscopic objects, one discovers that the wavelengths of even the smallest objects visible to the naked eye are too small for the wavelike characteristics of these objects to be significant in any real situation. For the electron, however, the wavelength is large enough to be measurable. Diffraction experiments have been performed using electrons, demonstrating conclusively that they have wave properties.

It is contrary to our intuition that electrons might behave as waves. The repercussions of this notion are that the electron does not have a definite size, but is spread out over a region in space. We are more comfortable with the thought of the electron being a microscopic particle, moving around in an orbit near the nucleus of an atom. As with light, however, we do not abandon the particle model for electrons; rather, we employ mathematical equations arising from both particle and wave models. For quantum chemical calculations, the wave model turns out to be more useful.

The Heisenberg Uncertainty Principle

One consequence of the wave nature of matter is that the position and momentum of small objects are not well known, as they would be for a particle model. In some circumstances, a wave may be confined to a very narrow region in space; however, there is still some uncertainty as to its position. Additionally, the value of the momentum of a quantum object is often not known precisely. In 1927 Werner Heisenberg showed that the product of the uncertainty in position and the uncertainty in momentum is greater than or equal to a certain constant (Planck's constant divided by 4π). This constant is very small; accordingly, quantum mechanical uncertainty in position and momentum of objects that are large enough to see is not noticed experimentally. For electrons, however, quantum mechanical uncertainties in position and momentum are important considerations in interpreting both theoretical models and experimental results. The relationship between the uncertainties in position and momentum is known as the Heisenberg Uncertainty Principle. It tells us that the more we know about the position of a small object, such as an electron, the less we know about its momentum (and vice versa).

Calculating the Wavefunction

For a scientist, knowing that matter behaves as a wave is useful only if one knows something about that wave. The wavefunction is a mathematical function describing the wave. For example, $y(x) = A \sin(kx)$ might be the wave-function for a one-dimensional wave, which exists along the x -axis. Matter waves are three-dimensional; the relevant wavefunction depends on the x, y, and z coordinates of the system being studied (and sometimes on time as well). We conventionally label the wavefunction for a three-dimensional object as $\psi(x, y, z)$. In 1926 Erwin Schrödinger introduced a mathematical equation whereby, if one knows the potential energy acting on an object, one can calculate the wavefunction for that object. Heisenberg had already introduced a mathematical formalism for performing quantum mechanics calculations, without explicitly including the concept of waves. It was later shown that, although the approaches of Schrödinger and Heisenberg looked very different, they made exactly the same predictions. In practice, the Schrödinger formalism is more useful for explaining the problem being studied, and the Heisenberg methodology allows for more straightforward computation. Accordingly, a mixture of the two approaches is typically used in modern quantum chemistry. Once we know the wavefunction of the atom or molecule under study, we can calculate the properties of that atom or molecule.

Quantum Mechanics of Atoms

An exact solution for Schrödinger's wave equation can be obtained for the hydrogen atom; however, for larger atoms and molecules (which contain more than one electron), Schrödinger's equation can be solved only approximately. Although this may sound so restrictive as to make the equation useless, there are well-established approaches that allow for practical and accurate calculations on atoms and molecules. This is done by making some assumptions about larger systems based upon the hydrogen atom.

When the Schrödinger equation is solved for the hydrogen atom, the resulting wavefunctions for the various possible energies that the atom can have are used to determine atomic orbitals . An orbital is a region in space where an electron is most likely to be found. For example, the lowest-energy wavefunction for a hydrogen atom is the so-called 1s orbital , which is a spherical region in space surrounding the nucleus. For some higher-energy states, the orbitals are not necessarily spherical in shape.

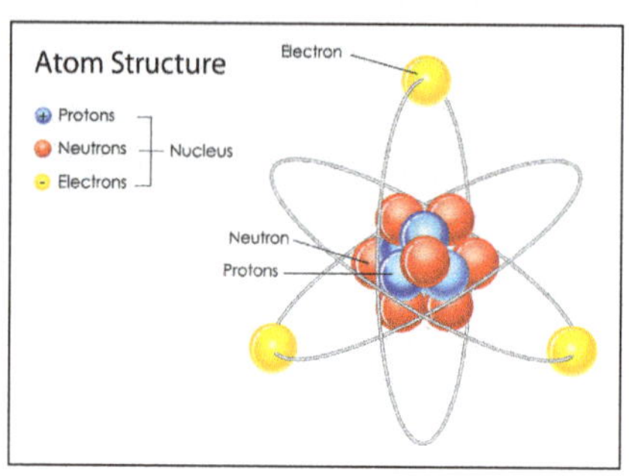

For atoms larger than hydrogen, one assumes that the orbitals occupied by the electrons have the same shape as the hydrogen orbitals but are differing in size and energy. The energies corresponding to these orbitals may be found by solving an approximate version of Schrödinger's equation. These atomic orbitals, in turn, may be used as the building blocks to the electronic behavior in molecules. As it happens, two electrons may share an atomic orbital; we say that these electrons are paired. Chemists have developed a system of rules for determining which orbitals are occupied in which atoms; calculations can then be done to determine the energies of the electrons in the atoms.

Quantum Mechanics of Molecules

Molecules are held together by covalent bonds. The simplest definition of a covalent bond is a shared pair of electrons. There are two basic approaches to modeling covalent bonds in molecules: the valence bond model and the molecular orbital model. In the valence bond model, we think of atomic orbitals on each of two atoms combining to form a bond orbital, with one electron from each atom occupying this orbital. Both the bond orbital and the electron pair now "belong" to both of the atoms. This sharing of electrons brings about a lowering in the energy, which makes the formation of molecules from atoms an energetically favorable process. The valence bond model gives the simplest quantum mechanical picture of chemical bonding, but it is not the best method for accurate calculations on molecules containing more than two atoms.

Molecular orbital theory differs from valence bond theory in that it does not require the electrons involved in a bond to be localized between two of the atoms in a molecule. Instead, the electron occupies a molecular orbital, which may be spread out over the entire molecule. As in the valence bond approach, the molecular orbital is formed by adding up contributions from the atomic orbitals on the atoms that make up the molecule. This approach, which does not explicitly model bonds as existing between two atoms, is somewhat less appealing to the intuition than the valence bond approach. However, molecular orbital calculations typically yield better predictions of molecular structure and properties than valence bond methods. Accordingly, most commercially available quantum chemistry software packages rely on molecular orbital methods to perform calculations.

Electronic Structure

Electronic structure is the state of motion of electrons in an electrostatic field created by stationary nuclei. The terms encompass both the wave functions of the electrons and the energies associated with them. Electrons in an atom are grouped around the nucleus into shells.

Shell (electron): A grouping of electrons in an atom according to energy. The farther a shell is from the nucleus, the larger it is, the more electrons it can hold, and the higher the energies of those electrons.

The first shell (closest to the nucleus) can hold two electrons. The second shell can hold 8 electrons. The third shell can hold 32 electrons.

Within the shells, electrons are further grouped into subshells of four different types, identified as s, p, d, and f in order of increasing energy. The first shell has only an s subshell; the second shell has an s and a p subshell; the third shell has s, p, and d subshells, and the fourth has s, p, d and f subshells. The number of subshells is equal to the shell number. A specific subshell is symbolized by writing the number of the shell, followed by the letter for the subshell.

Subshell (electron): A grouping of electrons in a shell according to the shape of the region of space they occupy.

Within each subshell, electrons are grouped into orbitals, regions of space within an atom where the specific electrons are most likely to be found. Within each subshell, electrons are grouped into orbitals, regions of space within an atom where the specific electrons are most likely to be found. Each orbital holds two electrons which differ in a property known as spin.

Orbital: A region of space within an atom where an electron in a given subshell can be found.

$$\begin{array}{lcccc} \text{Shell number:} & 1 & 2 & 3 & 4 \\ \text{Subshell designation:} & s & \overbrace{s,p} & \overbrace{s,p,d} & \overbrace{s,p,d,f} \\ \text{Number of orbitals:} & 1 & 1,3 & 1,3,5 & 1,3,5,7 \end{array}$$

Any orbital can hold a maximum of 2 electrons with opposite spin. The first shell has one 1s orbital and holds 2 electrons. The second shell holds 8 electrons; 2 in a 2s orbital and 6 in three 2p orbitals. The third shell holds 18 electrons; 2 in a 3s orbital; 6 in three 3p orbitals; and 10 in five 3d orbitals. The fourth shell holds 32 electrons; 2 in a 4s orbital; 6 in three 4p orbitals; 10 in five 4d orbitals; and 14 in seven 4f orbitals.

Table: Electron distribution in atoms.

Shell number	1	2	3	4
Subshell Designation	s	S,p	S,p,d	S, p, d, f
Number of or bitals	1	1,3	1,3,5	1,3,,5,7
Number of electrons	2	2,6	2,6,10	2,6,10,14
Total electron capacity	2	8	18	32

Electron Configurations

The exact arrangement of electrons in an atom's shells and subshells is the atom's electron configuration. It can be predicted by applying three rules.

Electron Configuration: The specific arrangement of electrons in an atom's shell and subshells.

1. Rule 1: Electrons occupy the lowest energy orbitals available. This is complicated by "crossover" of energies above the 3p level.
2. Rule 2: Each orbital can hold only two electrons, which must be of opposite spin.
3. Rule 3: Two or more orbitals with the same energy are each half-filled by one electron before any one orbital is completely filled by addition of the second electron.

Below is a simple scheme to help remember the order in which the orbitals are filled.

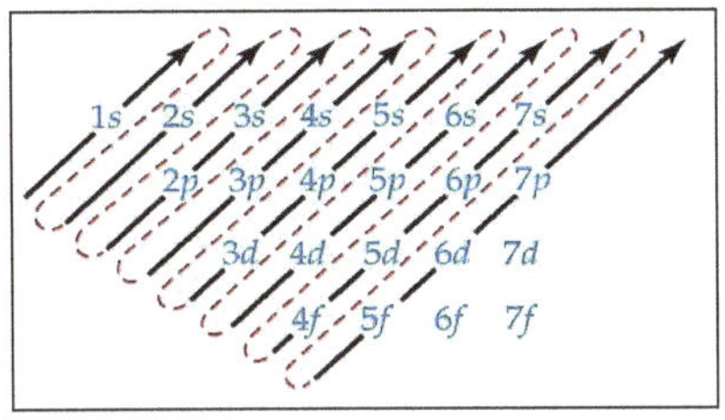

The number of electrons in each subshell is indicated by a superscript.

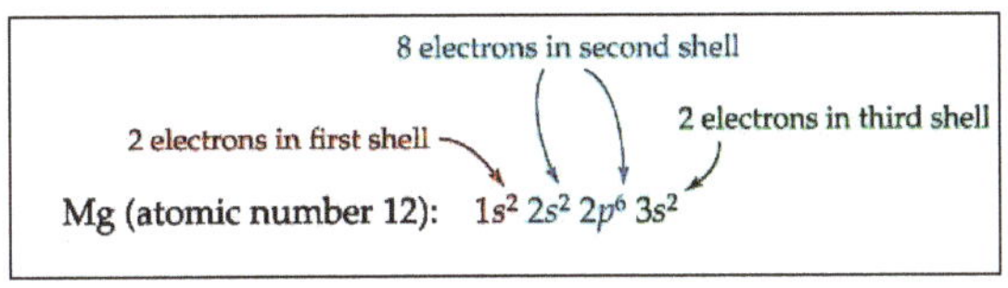

These are the electron configurations for B – N in which the 2p shell begins to fill.

B $\quad 1s^2 2s^2 2p^1$ or $\underset{1s^2}{\underline{\uparrow\downarrow}}\ \underset{2s^2}{\underline{\uparrow\downarrow}}\ \underbrace{\uparrow\ \ \ }_{2p^1}$ or $[\mathrm{He}]2s^2 2p^1$

C $\quad 1s^2 2s^2 2p^2$ or $\underset{1s^2}{\underline{\uparrow\downarrow}}\ \underset{2s^2}{\underline{\uparrow\downarrow}}\ \underbrace{\uparrow\uparrow\ }_{2p^2}$ or $[\mathrm{He}]2s^2 2p^2$

N $\quad 1s^2 2s^2 2p^3$ or $\underset{1s^2}{\underline{\uparrow\downarrow}}\ \underset{2s^2}{\underline{\uparrow\downarrow}}\ \underbrace{\uparrow\uparrow\uparrow}_{2p^3}$ or $[\mathrm{He}]2s^2 2p^3$

These are the electron configurations for O – Ne in which the 2p shell is completed.

O $\quad 1s^2 2s^2 2p^4$ or $\underset{1s^2}{\underline{\uparrow\downarrow}}\ \underset{2s^2}{\underline{\uparrow\downarrow}}\ \underbrace{\uparrow\downarrow\uparrow\uparrow}_{2p^4}$ or $[\mathrm{He}]2s^2 2p^4$

F $\quad 1s^2 2s^2 2p^5$ or $\underset{1s^2}{\underline{\uparrow\downarrow}}\ \underset{2s^2}{\underline{\uparrow\downarrow}}\ \underbrace{\uparrow\downarrow\uparrow\downarrow\uparrow}_{2p^5}$ or $[\mathrm{He}]2s^2 2p^5$

Ne $\quad 1s^2 2s^2 2p^6$ or $\underset{1s^2}{\underline{\uparrow\downarrow}}\ \underset{2s^2}{\underline{\uparrow\downarrow}}\ \underbrace{\uparrow\downarrow\uparrow\downarrow\uparrow\downarrow}_{2p^6}$

Electronic Structure Theory

Electronic Structure Theory describes the motions of electrons in atoms or molecules. Generally this is done in the context of the Born-Oppenheimer Approximation, which says that electrons are so much lighter (and therefore faster) than nuclei that they will find their optimal distribution for any given nuclear configuration. The electronic energy at each nuclear configuration is the potential energy that the nuclei feel, so solving the electronic problem for a range of nuclear configurations gives the potential energy surface.

Because the electrons are so small, one needs to use quantum mechanics to solve for their motion. Quantum mechanics tells us that the electrons will not be localized at particular points in space, but they are best thought of as "matter waves" which can interfere. The probability of finding a single electron at a given point in space is given by $\Psi^*(x)\Psi(x)$ for its wavefunction Ψ at the point x. The wavefunction can be determined by solving the time-independent Schrödinger equation $\hat{H}\Psi = E\Psi$. If the problem is time-dependent, then the time-dependent Schrödinger equation $i\hbar\frac{\partial\Psi}{\partial t} = \hat{H}\Psi$ must be used instead; otherwise, the solutions to the time-independent problem are also solutions to the time-dependent problem when they are multiplied by the energy dependent phase factor $e^{-iEt/\hbar}$. Since we have fixed the nuclei under the Born-Oppenheimer approximation, we solve for the nonrelativistic *electronic* Schrödinger equation:

$$\hat{H} = -\frac{\hbar^2}{2m}\sum_i \nabla_i^2 - \sum_A \frac{\hbar^2}{2M_A}\nabla_A^2 - \sum_{A,i}\frac{Z_A e^2}{4\pi \epsilon_0 r_{Ai}} + \sum_{A>B}\frac{Z_A Z_B e^2}{4\pi \epsilon_0 R_{AB}} + \sum_{i>j}\frac{e^2}{4\pi \epsilon_0 r_{ij}},$$

Where i, j refer to electrons and A, B refer to nuclei. In atomic units, this simplifies to:

$$\hat{H} = -\frac{1}{2}\sum_i \nabla_i^2 - \sum_A \frac{1}{2M_A}\nabla_A^2 - \sum_{A,i}\frac{Z_A}{r_{Ai}} + \sum_{A>B}\frac{Z_A Z_B}{R_{AB}} + \sum_{i>j}\frac{1}{r_{ij}}.$$

This Hamiltonian is appropriate as long as relativistic effects are not important for the system in question. Roughly speaking, relativistic effects are not generally considered important for atoms with atomic number below about 25 (Mn). For heavier atoms, the inner electrons are held more tightly to the nucleus and have velocities which increase as the atomic number increases; as these velocities approach the speed of light, relativistic effects become more important. There are various approaches for accounting for relativistic effects, but the most popular is to use relativistic effective core potentials (RECPs), often along with the standard nonrelativistic Hamiltonian.

Valence Bond Theory

Valence bond theory is most often used to describe bonding in organic molecules. In this model, bonds are considered to form from the overlap of two atomic orbitals on different atoms, each orbital containing a single electron. In looking at simple inorganic molecules such as molecular hydrogen (H_2) or hydrogen fluoride (HF), our present understanding of *s* and *p* atomic orbitals will suffice. In order to explain the bonding in organic molecules, however, we will need to introduce the concept of hybrid orbitals.

The Sigma Bond in the H_2 Molecule

The simplest case to consider is the hydrogen molecule, H_2. When we say that the two hydrogen nuclei share their electrons to form a covalent bond, what we mean in valence bond theory terms is that the two spherical 1*s* orbitals overlap, and contain two electrons with opposite spin.

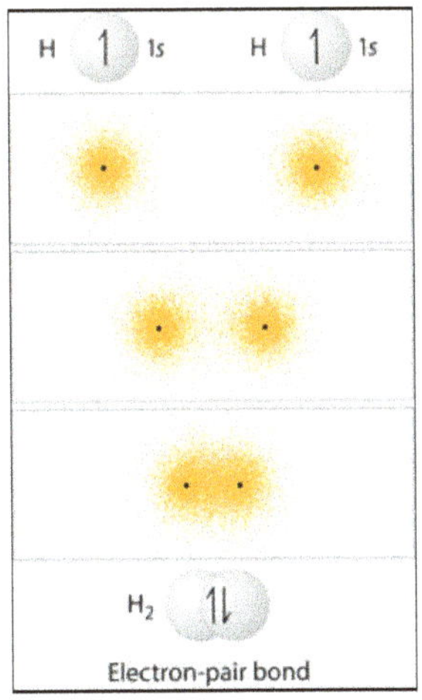

These two electrons are now attracted to the positive charge of *both* of the hydrogen nuclei, with the result that they serve as a sort of 'chemical glue' holding the two nuclei together.

How far apart are the two nuclei? If they are too far apart, their respective 1s orbitals cannot overlap, and thus no covalent bond can form - they are still just two separate hydrogen atoms. As they move closer and closer together, orbital overlap begins to occur, and a bond begins to form. This lowers the potential energy of the system, as new, *attractive* positive-negative electrostatic interactions become possible between the nucleus of one atom and the electron of the second.

But something else is happening at the same time: as the atoms get closer, the *repulsive* positive-positive interaction between the two nuclei also begins to increase.

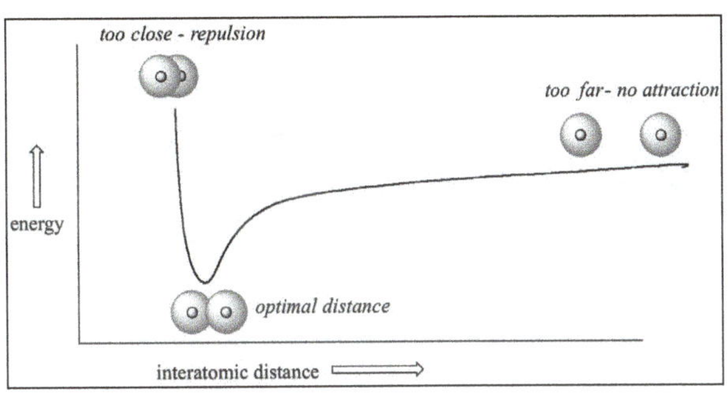

At first this repulsion is more than offset by the attraction between nuclei and electrons, but at a certain point, as the nuclei get even closer, the repulsive forces begin to overcome the attractive forces, and the potential energy of the system rises quickly. When the two nuclei are 'too close', we have an unstable, high-energy situation. There is a defined optimal distance between the nuclei in which the potential energy is at a minimum, meaning that the combined attractive and repulsive forces add up to the greatest overall attractive force. This optimal internuclear distance is the bond length. For the H_2 molecule, the distance is 74 pm (picometers, 10^{-12} meters). Likewise, the difference in potential energy between the lowest energy state (at the optimal internuclear distance) and the state where the two atoms are completely separated is called the bond dissociation energy, or, more simply, bond strength. For the hydrogen molecule, the H-H bond strength is equal to about 435 kJ/mol.

Every covalent bond in a given molecule has a characteristic length and strength. In general, the length of a typical carbon-carbon single bond in an organic molecule is about 150 pm, while carbon-carbon double bonds are about 130 pm, carbon-oxygen double bonds are about 120 pm, and carbon-hydrogen bonds are in the range of 100 to 110 pm. The strength of covalent bonds in organic molecules ranges from about 234 kJ/mol for a carbon-iodine bond (in thyroid hormone, for example), about 410 kJ/mole for a typical carbon-hydrogen bond, and up to over 800 kJ/mole for a carbon-carbon triple bond.

It is not accurate, however, to picture covalent bonds as rigid sticks of unchanging length - rather, it is better to picture them as *springs* which have a defined length when relaxed, but which can be compressed, extended, and bent.

One more characteristic of the covalent bond in H_2 is important to consider at this point. The two overlapping 1*s* orbitals can be visualized as two spherical balloons being pressed together. This means that the bond has cylindrical symmetry: if we were to take a cross-sectional plane of the bond at any point, it would form a circle. This type of bond is referred to as a σ (sigma) bond.

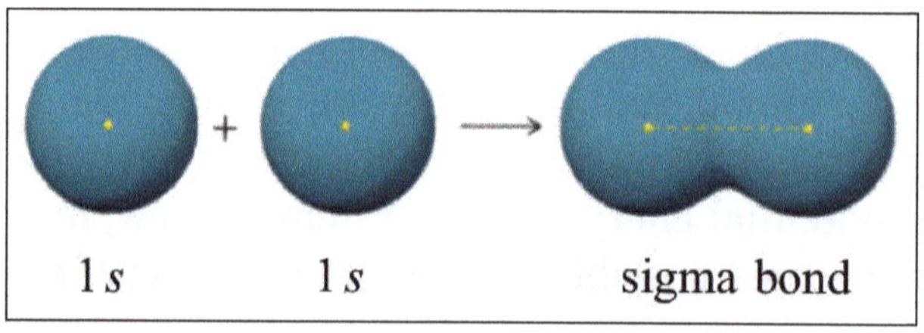

A sigma bond can be formed by overlap of an *s* atomic orbital with a *p* atomic orbital. Hydrogen fluoride (HF) is an example:

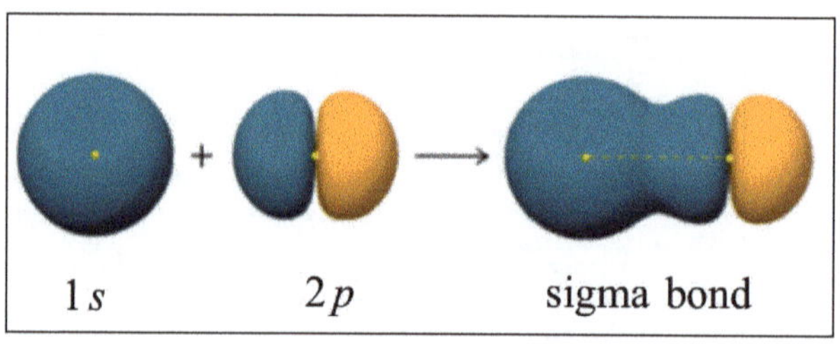

A sigma bond can also be formed by the overlap of two *p* orbitals. The covalent bond in molecular fluorine, F_2, is a sigma bond formed by the overlap of two half-filled 2*p* orbitals, one from each fluorine atom.

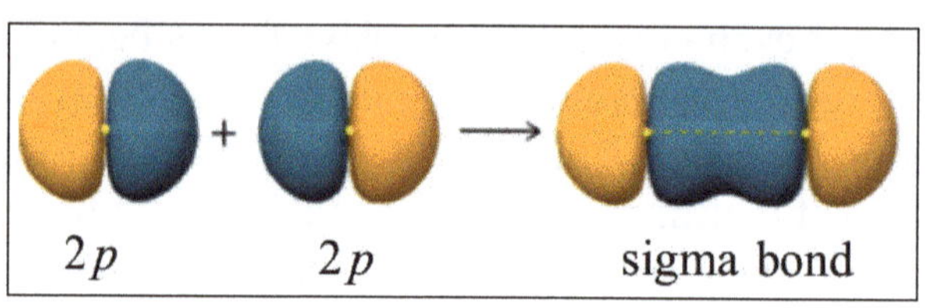

*sp*³ Hybrid Orbitals and Tetrahedral Bonding

Now let's look more carefully at bonding in organic molecules, starting with methane, CH_4. Recall the valence electron configuration of a carbon atom.

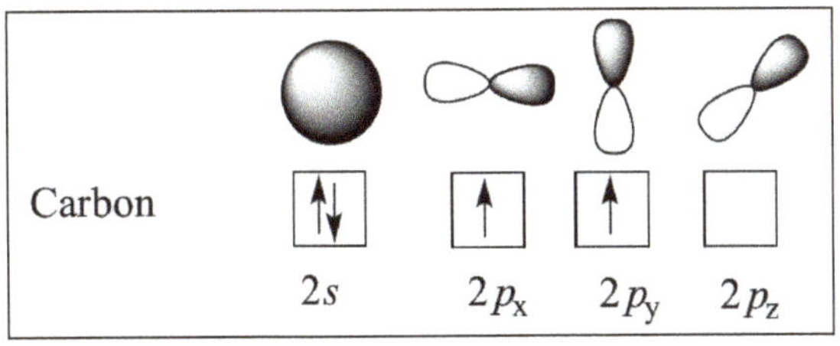

This picture is problematic when it comes to describing the *bonding* in methane. How does the carbon form four bonds if it has only two half-filled *p* orbitals available for bonding? A hint comes from the experimental observation that the four C-H bonds in methane are arranged with tetrahedral geometry about the central carbon, and that each bond has the same length and strength. In order to explain this observation, valence bond theory relies on a concept called orbital hybridization. In this picture, the four valence orbitals of the carbon (one 2*s* and three 2*p* orbitals) combine mathematically (remember: orbitals are described by wave equations) to form four equivalent hybrid orbitals, which are called sp^3orbitals because they are formed from mixing one *s* and three *p* orbitals. In the new electron configuration, each of the four valence electrons on the carbon occupies a single sp^3 orbital.

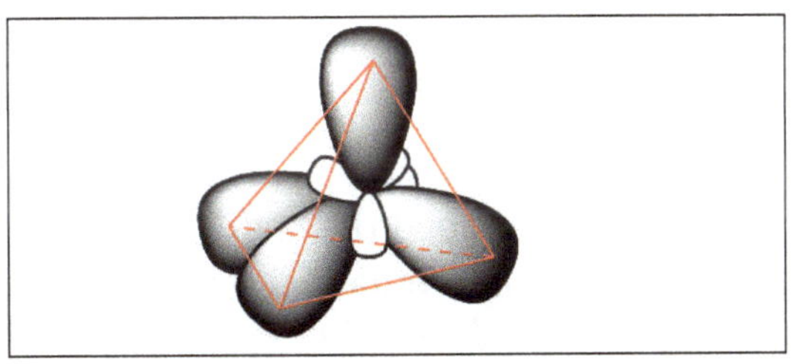

Four sp^3 orbitals in tetrahedral orientation.

This geometric arrangement makes perfect sense if you consider that it is precisely this angle that allows the four orbitals (and the electrons in them) to be as far apart from each other as possible. This is simply a restatement of the Valence Shell Electron Pair Repulsion (VSEPR) theory that you learned in General Chemistry: electron pairs (in orbitals) will arrange themselves in such a way as to remain as far apart as possible, due to negative-negative electrostatic repulsion.

Each C-H bond in methane, then, can be described as a sigma bond formed by overlap between a half-filled 1*s* orbital in a hydrogen atom and the larger lobe of one of the four half-filled sp^3 hybrid orbitals in the central carbon. The length of the carbon-hydrogen bonds in methane is 109 pm.

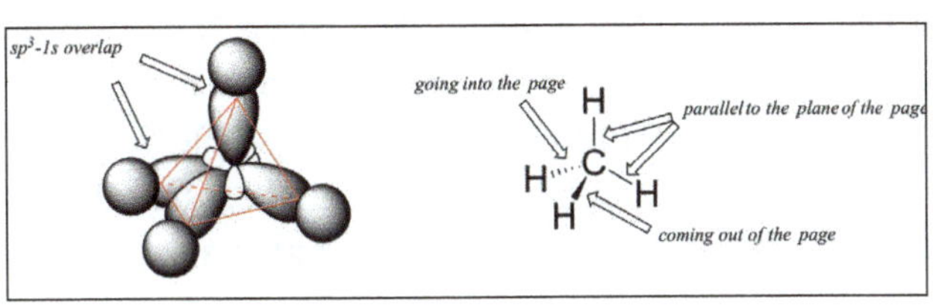

Tetrahedral sp^3 bonding in methane.

How does this bonding picture extend to compounds containing carbon-carbon bonds? In ethane (CH_3CH_3), both carbons are sp^3-hybridized, meaning that both have four bonds with tetrahedral geometry. The carbon-carbon bond, with a bond length of 154 pm, is formed by overlap of one sp^3

orbital from each of the carbons, while the six carbon-hydrogen bonds are formed from overlaps between the remaining sp^3 orbitals on the two carbons and the 1s orbitals of hydrogen atoms. All of these are sigma bonds.

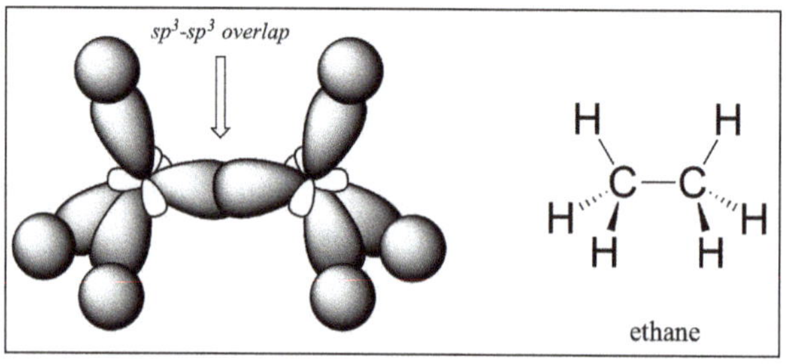

Because they are formed from the end-on-end overlap of two orbitals, *sigma bonds are free to rotate*. This means, in the case of ethane molecule, that the two methyl (CH_3) groups can be pictured as two wheels on an axle, each one able to rotate with respect to the other.

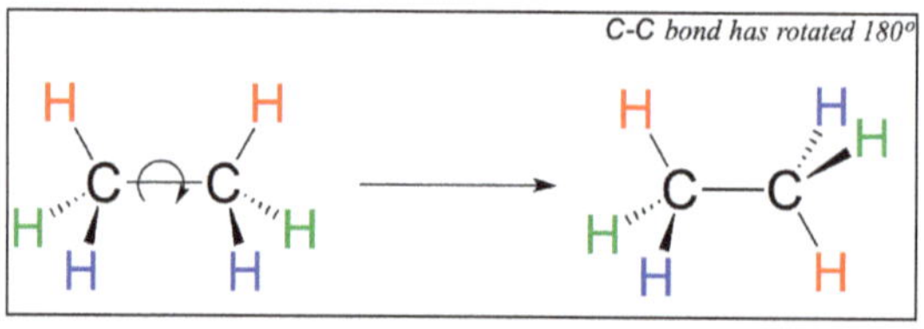

The sp^3 bonding picture is also used to described the bonding in amines, including ammonia, the simplest amine. Just like the carbon atom in methane, the central nitrogen in ammonia is sp^3-hybridized. With nitrogen, however, there are five rather than four valence electrons to account for, meaning that three of the four hybrid orbitals are half-filled and available for bonding, while the fourth is fully occupied by a nonbonding pair (lone pair) of electrons.

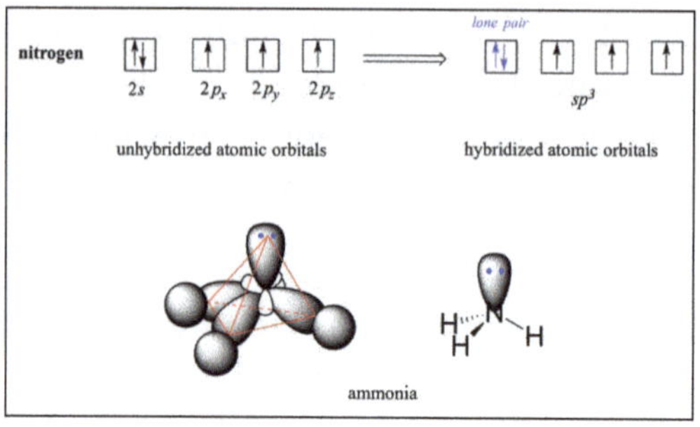

The bonding arrangement here is also tetrahedral: the three N-H bonds of ammonia can be pictured as forming the base of a trigonal pyramid, with the fourth orbital, containing the lone pair, forming the top of the pyramid. Recall from your study of VSEPR theory in General Chemistry that the lone pair, with its slightly greater repulsive effect, 'pushes' the three N-H s bonds away from the top of the pyramid, meaning that the H-N-H bond angles are slightly less than tetrahedral, at 107.3° rather than 109.5°.

VSEPR theory also predicts, accurately, that a water molecule is 'bent' at an angle of approximately 104.5°. The bonding in water results from overlap of two of the four sp^3 hybrid orbitals on oxygen with 1s orbitals on the two hydrogen atoms. The two nonbonding electron pairs on oxygen are located in the two remaining sp^3 orbitals.

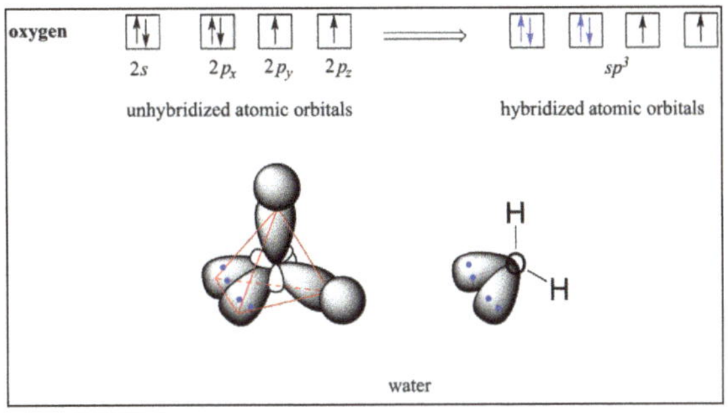

*sp*² and *sp* Hybrid Orbitals and pi Bonds

The valence bond theory, along with the hybrid orbital concept, does a very good job of describing double-bonded compounds such as ethene. Three experimentally observable characteristics of the ethene molecule need to be accounted for by a bonding model:

1. Ethene is a planar (flat) molecule.
2. Bond angles in ethene are approximately 120°, and the carbon-carbon bond length is 134 pm, significantly shorter than the 154 pm single carbon-carbon bond in ethane.
3. There is a significant barrier to rotation about the carbon-carbon double bond.

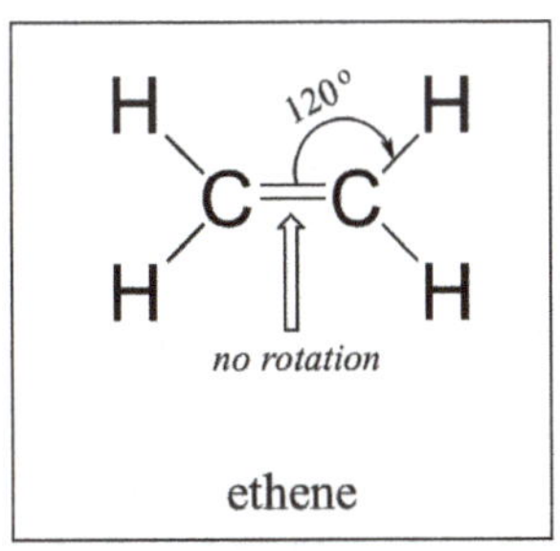

Clearly, these characteristics are not consistent with an sp^3 hybrid bonding picture for the two carbon atoms. Instead, the bonding in ethene is described by a model involving the participation of a different kind of hybrid orbital. Three atomic orbitals on each carbon – the 2*s*, $2p_x$ and $2p_y$ orbitals – combine to form three sp^2 hybrids, leaving the $2p_z$ orbital unhybridized.

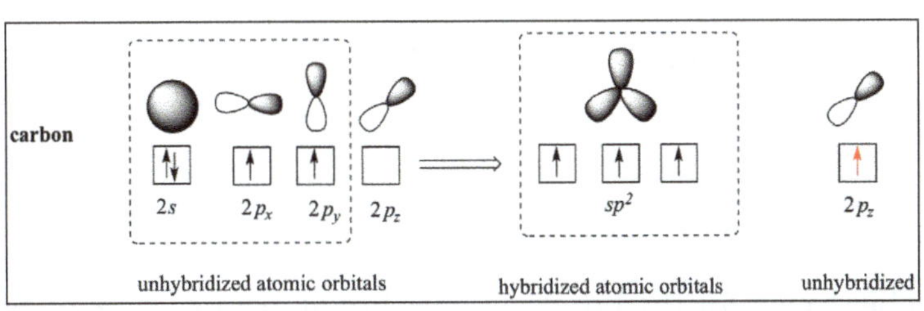

The three sp^2 hybrids are arranged with trigonal planar geometry, pointing to the three corners of an equilateral triangle, with angles of 120° between them. The unhybridized $2p_z$ orbital is *perpendicular* to this plane (in the next several figures, sp^2 orbitals and the sigma bonds to which they contribute are represented by lines and wedges; only the $2p_z$ orbitals are shown in the 'space-filling' mode).

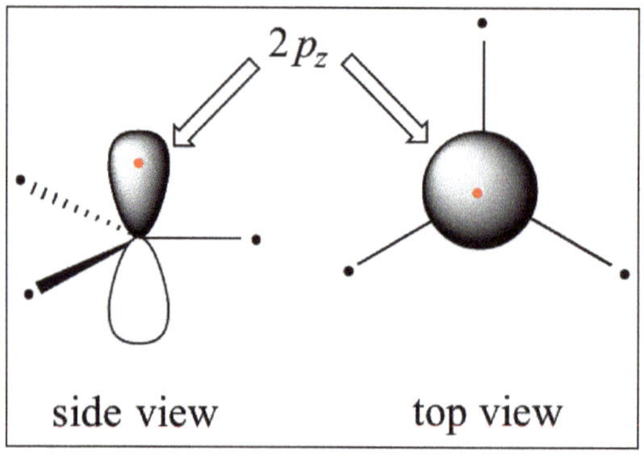

The carbon-carbon double bond in ethene consists of one sigma bond, formed by the overlap of two sp^2 orbitals, and a second bond, called a pi bond, which is formed by the *side-by-side* overlap of the two unhybridized $2p_z$ orbitals from each carbon.

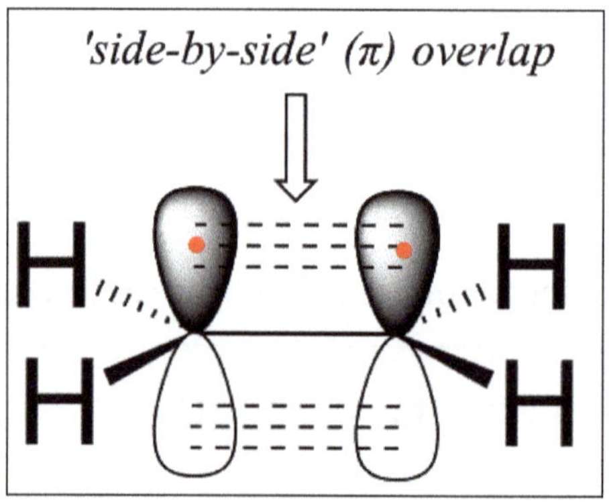

Unlike a sigma bond, a pi bond does *not* have cylindrical symmetry. If rotation about this bond were to occur, it would involve disrupting the side-by-side overlap between the two $2p_z$ orbitals that make up the pi bond. The presence of the pi bond thus 'locks' the six atoms of ethene into the same plane. This argument extends to larger alkene groups: in each case, six atoms lie in the same plane.

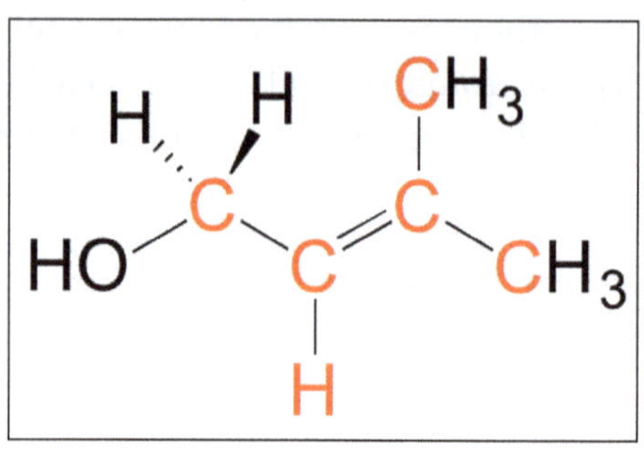

A similar picture can be drawn for the bonding in carbonyl groups, such as formaldehyde. In this molecule, the carbon is sp^2-hybridized, and we will assume that the oxygen atom is also sp^2 hybridized. The carbon has three sigma bonds: two are formed by overlap between sp^2 orbitals with $1s$ orbitals from hydrogen atoms, and the third sigma bond is formed by overlap between the remaining carbon sp^2 orbital and an sp^2 orbital on the oxygen. The two lone pairs on oxygen occupy its other two sp^2 orbitals.

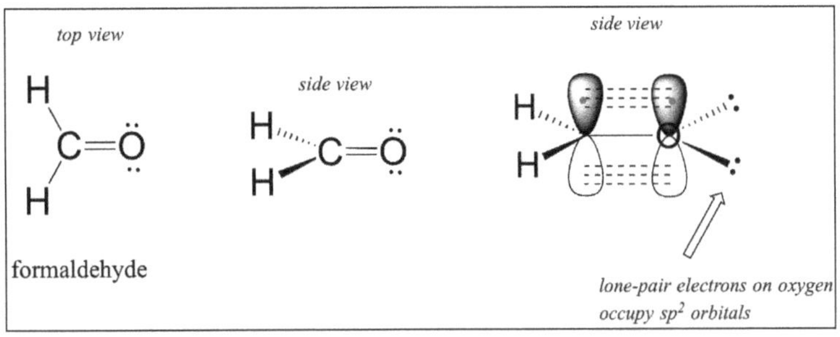

The pi bond is formed by side-by-side overlap of the unhybridized $2p_z$ orbitals on the carbon and the oxygen. Just like in alkenes, the $2p_z$ orbitals that form the pi bond are perpendicular to the plane formed by the sigma bonds.

Recall that carbocations are transient, high-energy species in which a carbon only has three bonds (rather than the usual four) and a positive formal charge. For now, though, the important thing to understand is that a carbocation can be described as an sp^2-hybridized carbon with an *empty p* orbital perpendicular to the plane of the sigma bonds.

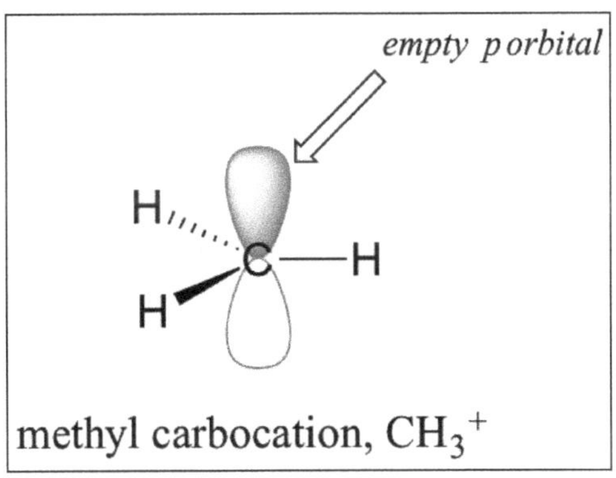

Finally, the hybrid orbital concept applies as well to triple-bonded groups, such as alkynes and nitriles. Consider, for example, the structure of ethyne (common name acetylene), the simplest alkyne.

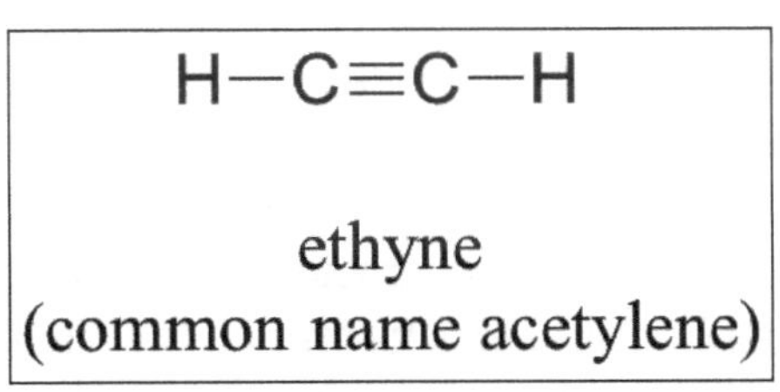

Both the VSEPR theory and experimental evidence tells us that the molecule is linear: all four atoms lie in a straight line. The carbon-carbon triple bond is only 120 pm long, shorter than the double bond in ethene, and is very strong, about 837 kJ/mol. In the hybrid orbital picture of acetylene, both carbons are *sp*-hybridized. In an *sp*-hybridized carbon, the 2*s*orbital combines with the $2p_x$ orbital to form two *sp* hybrid orbitals that are oriented at an angle of 180° with respect to each other (eg. along the x axis). The $2p_y$ and $2p_z$ orbitals remain unhybridized, and are oriented perpendicularly along the y and z axes, respectively.

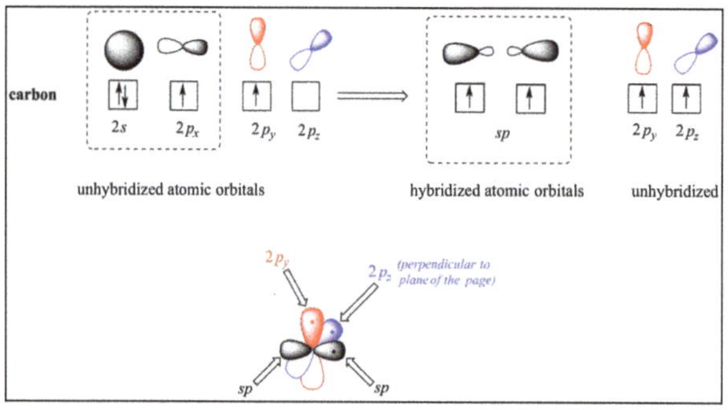

The carbon-carbon sigma bond, then, is formed by the overlap of one *sp* orbital from each of the carbons, while the two carbon-hydrogen sigma bonds are formed by the overlap of the second *sp* orbital on each carbon with a 1*s* orbital on hydrogen. Each carbon atom still has two half-filled $2p_y$ and $2p_z$ orbitals, which are perpendicular both to each other and to the line formed by the sigma bonds. These two perpendicular pairs of *p* orbitals form two pi bonds between the carbons, resulting in a triple bond overall (one sigma bond plus two pi bonds).

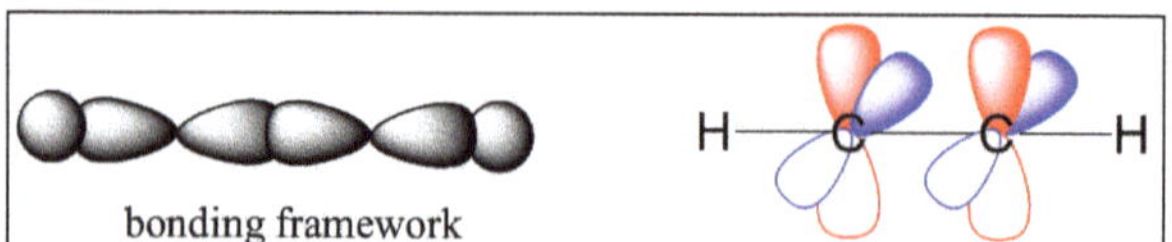

bonding framework in ethyne

π overlap in ethyne

The hybrid orbital concept nicely explains another experimental observation: single bonds adjacent to double and triple bonds are progressively shorter and stronger than single bonds adjacent to other single bonds. Consider for example, the carbon-carbon single bonds in propane, propene, and propyne.

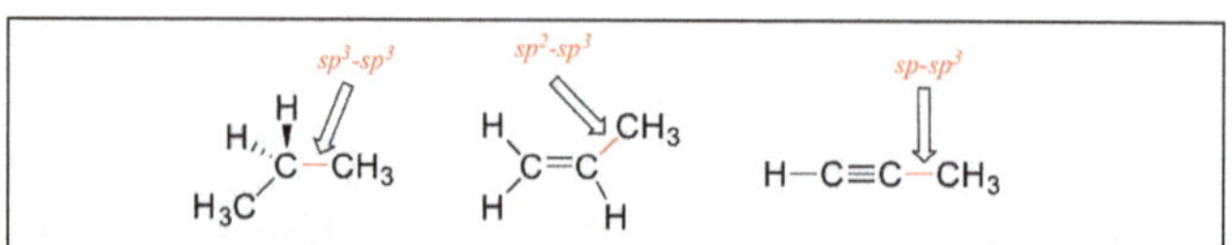

All three are single (sigma) bonds; the bond in propyne is shortest and strongest, while the bond in propane is longest and weakest. The explanation is relatively straightforward. An *sp* orbital is composed of one *s* orbital and one *p* orbital, and thus it has 50% *s* character and 50% *p* character. sp^2 orbitals, by comparison, have 33% *s* character and 67% *p* character, while sp^3 orbitals have 25% *s* character and 75% *p* character. Because of their spherical shape, 2*s* orbitals are smaller, and hold electrons closer and 'tighter' to the nucleus, compared to 2*p* orbitals. It follows that electrons in an *sp* orbital, with its greater *s* character, are closer to the nucleus than electrons in an sp^2 or sp^3 orbital. Consequently, bonds involving sp-sp^3 overlap (as in propyne) are shorter and stronger than bonds involving sp^2-sp^3 overlap (as in propene). Bonds involving sp^3-sp^3 overlap (as in propane) are the longest and weakest of the three.

Molecular Orbital Theory

The Valence Bond Theory fails to answer certain questions like why He_2 molecule does not exist and why O_2 is paramagnetic.

According to the Molecular Orbital Theory, individual atoms combine to form molecular orbitals. Thus the electrons of an atom are present in various atomic orbitals and are associated with several nuclei.

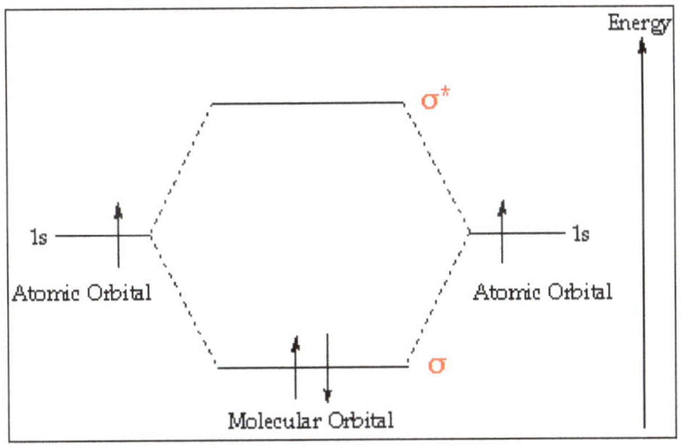

We know that we can consider electrons as either particle or wave nature. Therefore, we can describe an electron in an atom as occupying an atomic orbital, or by a wave function Ψ. These are solutions to the Schrodinger wave equation. Electrons in a molecule occupy molecular orbitals. We can obtain the wave function of a molecular orbital by the following methods:

- Linear combination of atomic orbitals (LCAO)
- United atom method

Linear Combination of Atomic Orbitals (LCAO)

As per this method, the formation of orbitals is because of Linear Combination (addition or subtraction) of atomic orbitals which combine to form the molecule. Consider two atoms A and B which have atomic orbitals described by the wave functions Ψ_A and Ψ

If the electron cloud of these two atoms overlaps, then we can obtain the wave function for the molecule by a linear combination of the atomic orbitals Ψ_A and Ψ_B. The below equation forms two molecular orbitals.

$$\Psi_{MO} = \Psi_A \pm \Psi_B$$

Bonding Molecular Orbitals

When the addition of wave function takes place, the type of molecular orbitals formed are Bonding Molecular Orbitals. We can represent them by $\Psi_{MO} = \Psi_A \pm \Psi_B$. They have lower energy than atomic orbitals involved.

Anti-Bonding Molecular Orbitals

When molecular orbital forms by the subtraction of wave function, the type of molecular orbitals formed are antibonding Molecular Orbitals. We can represent them as $\Psi_{MO} = \Psi_A - \Psi_B$. They have higher energy than atomic orbitals. Therefore, the combination of two atomic orbitals results in the formation of two molecular orbitals. They are the bonding molecular orbital (BMO) and the anti-bonding molecular orbital (ABMO).

Relative Energies of Molecular Orbitals:

- Bonding Molecular Orbitals (BMO) – Energy of Bonding Molecular Orbitals is less than that of Anti Bonding Molecular Orbitals. This is because of the increase in the attraction of both the nuclei for both the electron (of the combining atom).
- Anti-Bonding Molecular Orbitals (ABMO) – Energy of Anti Bonding Molecular Orbitals is higher than Bonding Molecular Orbitals. This is because the electrons tries to move away from the nuclei and are in a repulsive state.

Rules for Filling of Molecular Orbitals

We have to follow certain rules while filling up molecular orbitals with electrons in order to write correct molecular configurations. They are:

- Aufbau Principle – This principle states that those molecular orbitals which have the lowest energy are filled first.
- Pauli's Exclusion Principle – According to this principle, each molecular orbital can accommodate a maximum of two electrons having opposite spins.

Density Functional Theory

Density-functional theory is a set of theories in statistical mechanics that profit from the fact that the Helmholtz energy function of a system can be cast as a functional of the density. That is, the density (in its usual sense of particles per volume), which is a function of the position in inhomogeneous systems, uniquely defines the Helmholtz energy. By minimizing this Helmholtz energy one arrives at the true Helmholtz energy of the system and the equilibrium density function. The situation parallels the better known electronic density functional theory, in which the energy of a quantum system is shown to be a functional of the electronic density.

Starting from this fact, approximations are usually made in order to approach the true functional of a given system. An important division is made between *local* and *weighed* theories. In a local density theory in which the dependence is local, as exemplified by the (exact) Helmholtz energy of an ideal system:

$$A_{id} = k_B T \int dr \rho(r)[\log \rho(r) - 1 - U(r)],$$

where U(r) is an external potential. It is an easy exercise to show that Boltzmann's barometric law follows from minimization. An example of a weighed density theory would be the (also exact) excess Helmholtz energy for a system of 1-dimensional hard rods:

$$A_{ex} = -k_B T \int dz \rho(z) \log[1 - t(z)]$$

Where $t(z) = \int_{z-\sigma}^{z} dy \rho(y),$ precisely an average of the density over the length of the hard rods, σ. "Excess" means "over ideal", i.e., it is the total $A=A_{id}+A_{ex}$ that is to be minimized.

Chemical Dynamics

Chemical dynamics is a field in which scientists study the rates and mechanisms of chemical reactions. It also involves the study of how energy is transferred among molecules as they undergo collisions in gas-phase or condensed-phase environments. Therefore, the experimental and theoretical tools used to probe chemical dynamics must be capable of monitoring the chemical identity and energy content (i.e., electronic, vibrational, and rotational state populations) of the reacting species. Moreover, because the rates of chemical reactions and energy transfer are of utmost importance, these tools must be capable of doing so on time scales over which these processes, which are often very fast, take place. Let us begin by examining many of the most commonly employed theoretical models for simulating and understanding the processes of chemical dynamics.

Adiabatic and Nonadiabatic Dynamics

The BO approximation never explicitly addresses electronic or nuclear dynamics, but neglecting the nuclear kinetic energy to obtain potential energy surfaces has implicit dynamical consequences. As we discussed for our NaCl example, moving the neutral atoms together slowly allows electrons to completely equilibrate about each forward step, resulting in propagation on the adiabatic ground state. This is the essence of the adiabatic approximation. If you prepare the system in Ψ_α, an eigenstate of H at the initial time t_0, and propagate slowly enough, that Ψ_α will evolve as an eigenstate for all times:

$$H(t)\Psi_\alpha(t) = E_\alpha(t)\Psi_\alpha(t)$$

Equivalently this means that the nth eigenfunction of $H(t_0)$ will also be the nth eigenfunction of H(t). In this limit, there are no transitions between BO surfaces, and the dynamics only reflect the phases acquired from the evolving system. That is the time propagator can be expressed as:

$$U(t,t_0)_{adiabatic} = \sum_\alpha |\alpha\rangle\langle\alpha| \exp\left(-\frac{i}{\hbar}\int_{t_0}^{t} dt' E_\alpha(t')\right)$$

In the opposite limit, we also know that if the atoms were incident on each other so fast (with such high kinetic energy) that the electron did not have time to transfer at the crossing, that the system would pass smoothly through the crossing along the diabatic surface. In fact it is expected that the atoms would collide and recoil. This implies that there is an intermediate regime in which the velocity of the system is such that the system will split and follow both surfaces to some degree.

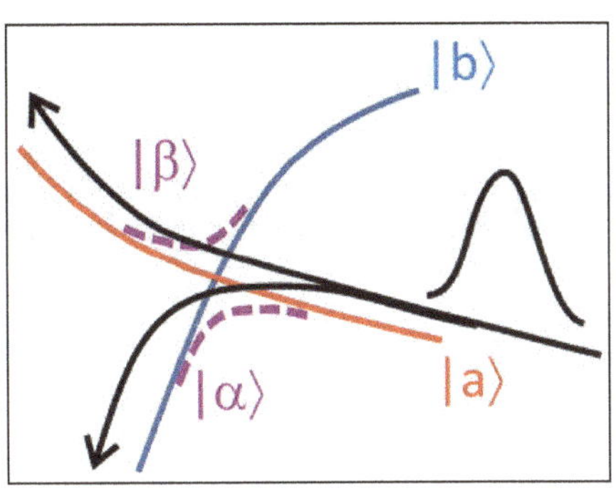

In a more general sense, we would like to understand the criteria for adiabaticity that enable a time-scale separation between the fast and slow degrees of freedom. Speaking qualitatively about any time-dependent interaction between quantum mechanical states, the time-scale that separates the fast and slow propagation regimes is determined by the strength of coupling between those states. We know that two coupled states exchange amplitude at a rate dictated by the Rabi frequency Ω_R, which in turn depends on the energy splitting and coupling of the states. For systems in which there is significant nonperturbative transfer of population between two states a and b, the time scale over which this can occur is approximately $\Delta t \approx 1/\Omega_R \approx \hbar/V_{ab}$. This is not precise, but is provides a reasonable starting point for discussing "slow" versus "fast". "Slow" in an adiabatic sense would mean that a time-dependent interaction act on the system over a period such that $\Delta t \ll \hbar/V_{ab}$. In the case of our NaCl example, we would be concerned with the time scale over which the atoms pass through the crossing region between diabatic states, which is determined by the incident velocity between atoms.

Adiabaticity Criterion

Let's investigate these issues by looking more carefully at the adiabatic approximation. Since the adiabatic states ($\Psi_a(t) \equiv |\alpha\rangle$ are orthogonal for all times, we can evaluate the time propagator as:

$$U(t) = \sum_{\alpha} e^{-\frac{i}{\hbar}\int_0^t E_\alpha(t)dt'} |\alpha\rangle\langle\alpha|$$

and the time-dependent wave function is:

$$\Psi(t) = \sum_{\alpha} b_\alpha(t) e^{-\frac{i}{\hbar}\int_0^t E_\alpha(t)dt'} |\alpha\rangle$$

Although these are adiabatic states we recognize that the expansion coefficients can be timedependent in the general case. So, we would like to investigate the factors that govern this time-dependence. To make the notation more compact, let's define the time-rate of change of the eigenfunction as:

$$|\dot{\alpha}\rangle = \frac{\partial}{\partial t}|\Psi\alpha(t)\rangle$$

If we substitute the general solution $\Psi(t) = \sum_{\alpha} b_\alpha(t)^{-\frac{i}{\hbar}\int_0^t E_\alpha(t)dt'} |\alpha\rangle$ into the TDSE, we get:

$$i\hbar\sum_{\alpha}\left(\dot{b}_\alpha|\alpha\rangle + b_\alpha|\dot{\alpha}\rangle - \frac{i}{\hbar}E_\alpha b_\alpha|\alpha\rangle\right)e^{-\frac{i}{\hbar}\int_0^t E_\alpha(t)dt'} = \sum_{\alpha}(b_\alpha E_\alpha|\alpha\rangle e^{-\frac{i}{\hbar}\int_0^t E_\alpha(t)dt'}$$

Note, the third term on the left hand side equals the right hand term. Acting on both sides from the left with $\langle\beta|$ leads to:

$$-\dot{b}_\beta e^{-\frac{i}{\hbar}\int_0^t E_\beta(t)dt'} = \sum_{\alpha} b_\alpha \langle\beta||\dot{\alpha}\rangle e^{-\frac{i}{\hbar}\int_0^t E_\alpha(t)dt'}$$

We can break up the terms in the summation into one for the target state $|\beta\rangle$ and one for the remaining states.

$$-\dot{b}_\beta = b_\beta \langle \beta | \dot{\beta} \rangle + \sum_{\alpha \neq \beta} b_\alpha \langle \beta | \dot{\alpha} \rangle \exp\left[-\frac{i}{\hbar} \int_0^t dt' E_{\alpha\beta}(t') \right]$$

Where,

$$E_{\alpha\beta}(t) = E_\alpha(t) - E_\beta(t).$$

The adiabatic approximation applies when we can neglect the summation in $-\dot{b}_\beta = b_\beta \langle \beta | \dot{\beta} \rangle + \sum_{\alpha \neq \beta} b_\alpha \langle \beta | \dot{\alpha} \rangle \exp\left[-\frac{i}{\hbar} \int_0^t dt' E_{\alpha\beta}(t') \right]$, or equivalently when $\langle \beta | \dot{\alpha} \rangle \ll \langle \beta | \dot{\beta} \rangle$ for all $|\alpha\rangle$. In that case, the system propagates on the adiabatic state $|\beta\rangle$ independent of the other states: $\dot{b}_\beta = -b_\beta \langle \beta | \dot{\beta} \rangle$. The evolution of the coefficients is,

$$\begin{aligned} b_\beta(t) &= b_\beta(0) \exp\left[-\int_0^t \langle \beta(t') | \dot{\beta}(t') \rangle dt' \right] \\ &\approx b_\beta(0) \exp\left[\frac{i}{\hbar} \int_0^t E_\beta(t') dt' \right] \end{aligned}$$

Here we note that in the adiabatic approximation:

$$E_\beta(t) = \langle \beta(t) | H(t) | \beta(t) \rangle.$$

Equation $\approx b_\beta(0) \exp\left[\frac{i}{\hbar} \int_0^t E_\beta(t') dt' \right]$ indicates that in the adiabatic approximation the population in the states never changes, only their phase. The second term on the right in equation $-\dot{b}_\beta = b_\beta \langle \beta | \dot{\beta} \rangle + \sum_{\alpha \neq \beta} b_\alpha \langle \beta | \dot{\alpha} \rangle \exp\left[-\frac{i}{\hbar} \int_0^t dt' E_{\alpha\beta}(t') \right]$ describes the nonadiabatic effects, and the overlap integral,

$$\langle \beta | \dot{\alpha} \rangle = \left\langle \Psi_\beta \middle| \frac{\partial \Psi_\alpha}{\partial t} \right\rangle$$

determines the magnitude of this effect. $\langle \beta | \dot{\alpha} \rangle$ is known as the *nonadiabatic coupling* (even though it refers to couplings between adiabatic surfaces), or as the geometrical phase. Note the parallels here to the expression for the nonadiabatic coupling in evaluating the validity of the Born Oppenheimer approximation, however, here the gradient of the wavefunction is evaluated in time rather than the nuclear position. It would appear that we can make some connections between these two results by linking the gradient variables through the momentum or velocity of the particles involved.

So, when can we neglect the nonadiabatic effects? We can obtain an expression for the nonadiabatic coupling by expanding:

$$\frac{\partial}{\partial t}\left[H|\alpha\rangle = E_\alpha|\alpha\rangle\right]$$

and acting from the left with ⟨β|, which for α≠β leads to:

$$\langle\beta|\dot{\alpha}\rangle = \frac{\langle\beta|\dot{H}|\alpha\rangle}{E_\alpha - E_\beta}$$

For adiabatic dynamics to hold $\langle\beta|\dot{\alpha}\rangle \ll \langle\beta|\dot{\beta}\rangle$ and so we can say,

$$\frac{\langle\beta|\dot{H}|\alpha\rangle}{E_\alpha - E_\beta} \ll -\frac{i}{\hbar}E_\beta$$

So how accurate is the adiabatic approximation for a finite time-period over which the systems propagates? We can evaluate equation $-\dot{b}_\beta = b_\beta\langle\beta|\dot{\beta}\rangle + \sum_{\alpha\neq\beta} b_\alpha\langle\beta|\dot{\alpha}\rangle \exp\left[-\frac{i}{\hbar}\int_0^t dt' E_{\alpha\beta}(t')\right]$, assuming that the system is prepared in state |α⟩ and that the occupation of this state never varies much from one. Then the occupation of any other state can be obtained by integrating over a period:

$$\begin{aligned}\dot{b}_\beta &= \langle\beta|\dot{\alpha}\rangle \exp\left[-\frac{i}{\hbar}\int_0^\tau dt' E_{\alpha\beta}(t')\right]\\ b_\beta &\approx i\hbar\frac{\langle\beta|\dot{H}|\alpha\rangle}{E_{\alpha\beta}^2}\left\{\exp\left[-\frac{i}{\hbar}E_{\alpha\beta\tau}\right]-1\right\}\\ &= 2\hbar\frac{\langle\beta|\dot{H}|\alpha\rangle}{E_{\alpha\beta}^2} e^{-\frac{i}{\hbar}E_{\alpha\beta\tau}} \sin\left(\frac{E_{\alpha\beta\tau}}{2\hbar}\right)\end{aligned}$$

Here, we used:

$$e^{i\theta} - 1 = 2ie^{i\theta/2}\sin(\theta/2).$$

For $|b_\beta k<1|$, we expand the sin term and find:

$$\left\langle\Psi_\beta\left|\frac{\partial H}{\partial t}\right|\Psi_\alpha\right\rangle \ll E_{\alpha\beta/\tau}$$

This is the criterion for adiabatic dynamics, which can be seen to break down near adiabatic curve crossings where $E_{\alpha\beta}$=0E, regardless of how fast we propagate through the crossing. Even away from curve crossing, there is always the possibility that nuclear kinetic energies are such that (∂H/∂t) will be greater than or equal to the energy splitting between adiabatic states.

Klopman-Salem Equation

In the theory of chemical reactivity, the Klopman-Salem equation describes the energetic change that occurs when two species approach each other in the course of a reaction and begin to interact, as their associated molecular orbitals begin to overlap with each other and atoms bearing partial charges begin to experience attractive or repulsive electrostatic forces. First described independently by Gilles Klopman and Lionel Salem in 1968, this relationship provides a mathematical basis for the key assumptions of frontier molecular orbital theory (i.e., theory of HOMO-LUMO interactions) and hard soft acid base (HSAB) theory. Conceptually, it highlights the importance of considering both electrostatic interactions and orbital interactions (and weighing the relative significance of each) when rationalizing the selectivity or reactivity of a chemical process.

Formulation and Interpretation

In modern form, the Klopman-Salem equation is commonly given as:

$$\Delta E = \Big(-\sum_{a,b}(q_a + q_b)\beta_{ab}S_{ab}\Big) + \Big(\sum_{k<\ell}\frac{Q_kQ_\ell}{\varepsilon R_{k\ell}}\Big) + \Big(\sum_{r}^{\text{occ.}}\sum_{s}^{\text{unocc.}} - \sum_{s}^{\text{occ.}}\sum_{r}^{\text{unocc.}}\frac{2(\sum_{a,b} c_{ra}c_{sb}\beta_{ab})^2}{E_r - E_s}\Big),$$

Where,

- q_a is the electron population in atomic orbital a,
- β_{ab}, S_{ab} are the resonance and overlap integrals for the interaction of atomic orbitals a and b,
- Q_k is the total charge on atom k,
- ε is the local dielectric constant,
- $R_{k\ell}$ is the distance between the nuclei of atoms k and l,
- c_{ra} is the coefficient of atomic orbital a in molecular orbital r, and
- E_r is the energy of molecular orbital r.

Broadly speaking, the first term describes the closed-shell repulsion of the occupied molecular orbitals of the reactants (four-electron filled-filled interactions, *steric effects*). The second term describes the Coulombic attraction or repulsion between the atoms of the reactants (ionic contribution, *electrostatic effects*). Finally, the third term accounts for all possible interactions between the occupied and unoccupied molecular orbitals of the reactants (two-electron filled-unfilled interactions, *stereoelectronic effects*). Although conceptually useful, the Klopman-Salem equation seldom serves as the basis for energetic analysis in modern quantum chemical calculations.

Because of the difference in MO energies appearing in the denominator of the third term, energetically close orbitals make the biggest contribution. Hence, approximately speaking, analysis can often be simplified by considering only the highest occupied and lowest unoccupied molecular orbitals of the reactants (the HOMO-LUMO interaction in frontier molecular orbital theory). The relative contributions of the second (ionic) and third (covalent) terms play an important role in justifying hard soft acid base theory (HSAB), with hard-hard interactions governed by the ionic term and soft-soft interactions governed by the covalent term.

Schrödinger's Equation

The Schrödinger equation is the fundamental equation of physics for describing quantum mechanical behavior. It is also often called the Schrödinger wave equation, and is a partial differential equation that describes how the wavefunction of a physical system evolves over time.

The Schrödinger equation has two 'forms', one in which time explicitly appears, and so describes how the wave function of a particle will evolve in time. In general, the wave function behaves like a wave, and so the equation is often referred to as the time dependent Schrödinger wave equation. The other is the equation in which the time dependence has been 'removed' and hence is known as the time independent Schrödinger equation and is found to describe, amongst other things, what the allowed energies are of the particle. These are not two separate, independent equations – the time independent equation can be derived readily from the time dependent equation (except if the potential is time dependent, a development we will not be discussing here). In the following we will describe how the first, time dependent equation can be 'derived', and in then how the second follows from the first.

Derivation of the Schrödinger Wave Equation

The Time Dependent Schrödinger Wave Equation

In the discussion of the particle in an infinite potential well, it was observed that the wave function of a particle of fixed energy E could most naturally be written as a linear combination of wave functions of the form:

$$\Psi(x,t) = Ae^{i(kx-\omega t)}$$

Representing a wave travelling in the positive x direction, and a corresponding wave travelling in the opposite direction, so giving rise to a standing wave, this being necessary in order to satisfy the boundary conditions. This corresponds intuitively to our classical notion of a particle bouncing back and forth between the walls of the potential well, which suggests that we adopt the wave function above as being the appropriate wave function for a free particle of momentum $p = \hbar k$ and energy $E = \hbar\omega$. With this in mind, we can then note that:

$$\frac{\partial^2\Psi}{\partial x^2} = -k^2\Psi$$

which can be written, using $E = p^2/2m = \hbar^2k^2/2m$:

$$-\frac{\hbar^2}{2m}\frac{\partial^2\Psi}{\partial x^2} = \frac{p^2}{2m}\Psi.$$

Similarly,

$$\frac{\partial \Psi}{\partial t} = -i\omega\Psi$$

which can be written, using $E = \hbar\omega$:

$$i\hbar\frac{\partial \Psi}{\partial t} = \hbar\omega\psi = E\Psi.$$

We now generalize this to the situation in which there is both a kinetic energy and a potential energy present, then $E = p^2/2m + V(x)$ so that:

$$E\Psi = \frac{p^2}{2m}\Psi + V(x)\Psi$$

where Ψ is now the wave function of a particle moving in the presence of a potential V (x). But if we assume that the results $-\frac{\hbar^2}{2m}\frac{\partial^2\Psi}{\partial x^2} = \frac{p^2}{2m}\Psi$ and $i\hbar\frac{\partial \Psi}{\partial t} = \hbar\omega\psi = E\Psi$ still apply in this case then we have,

$$-\frac{\hbar^2}{2m}\frac{\partial^2\psi}{\partial x^2} + V(x)\Psi = i\hbar\frac{\partial \psi}{\partial t}$$

which is the famous time dependent Schrödinger wave equation. It is setting up and solving this equation, then analyzing the physical contents of its solutions that form the basis of that branch of quantum mechanics known as wave mechanics.

Even though this equation does not look like the familiar wave equation that describes, for instance, waves on a stretched string, it is nevertheless referred to as a 'wave equation' as it can have solutions that represent waves propagating through space. We have seen an example of this: the harmonic wave function for a free particle of energy E and momentum p, i.e.,

$$\Psi(x,t) = Ae^{-i(px-Et)/\hbar}$$

is a solution of this equation with, as appropriate for a free particle, V (x) = 0. But this equation can have distinctly non-wave like solutions whose form depends, amongst other things, on the nature of the potential V (x) experienced by the particle.

In general, the solutions to the time dependent Schrödinger equation will describe the dynamical behaviour of the particle, in some sense similar to the way that Newton's equation F = ma describes the dynamics of a particle in classical physics. However, there is an important difference. By solving Newton's equation we can determine the position of a particle as a function of time, whereas by solving Schrödinger's equation, what we get is a wave function Ψ(x,t) which tells us (after we square the wave function) how the probability of finding the particle in some region in space varies as a function of time.

It is possible to proceed from here look at ways and means of solving the full, time dependent Schrödinger equation in all its glory, and look for the physical meaning of the solutions that are found. However this route, in a sense, bypasses much important physics contained in the Schrödinger equation which we can get at by asking much simpler questions. Perhaps the most important 'simpler question' to ask is this: what is the wave function for a particle of a given energy E? Curiously enough, to answer this question requires 'extracting' the time dependence from the time dependent Schrödinger equation. To see how this is done, and its consequences, we will turn our attention to the closely related time independent version of this equation.

The Time Independent Schrödinger Equation

We have seen what the wave function looks like for a free particle of energy E – one or the other of the harmonic wave functions – and we have seen what it looks like for the particle in an infinitely deep potential well though we did not obtain that result by solving the Schrödinger equation. But in both cases, the time dependence entered into the wave function via a complex exponential factor $\exp[-iEt/\hbar]$. This suggests that to 'extract' this time dependence we guess a solution to the Schrödinger wave equation of the form This suggests that to 'extract' this time dependence we guess a solution to the Schrödinger wave equation of the form:

$$\Psi(x,t) = \psi(x)e^{-iEt/\hbar}$$

i.e. where the space and the time dependence of the complete wave function are contained in separate factors. The idea now is to see if this guess enables us to derive an equation for ψ(x), the spatial part of the wave function.

If we substitute this trial solution into the Schrödinger wave equation, and make use of the meaning of partial derivatives, we get:

$$-\frac{\hbar^2}{2m}\frac{d^2\psi(x)}{dx^2}e^{-iEt/\hbar} + V(x)\psi(x)e^{-iEt/\hbar} = i\hbar. -iE/\hbar e^{-iEt/\hbar}\psi(x) = E\psi(x)e^{-iEt/\hbar}.$$

We now see that the factor $\exp[-iEt/\hbar]$ cancels from both sides of the equation, giving us,

$$-\frac{\hbar^2}{2m}\frac{d^2\psi(x)}{dx^2} + V(x)\psi(x) = E\psi(x)$$

If we rearrange the terms, we end up with:

$$\frac{\hbar^2}{2m}\frac{d^2\psi(x)}{dx^2} + (E - V(x))\psi(x) = 0$$

which is the time independent Schrödinger equation. We note here that the quantity E, which we have identified as the energy of the particle, is a free parameter in this equation. In other words, at no stage has any restriction been placed on the possible values for E. Thus, if we want to determine the wave function for a particle with some specific value of E that is moving in the presence of a potential V (x), all we have to do is to insert this value of E into the equation with the appropriate V (x), and solve for the corresponding wave function. In doing so, we find, perhaps not surprisingly, that for different choices of E we get different solutions for ψ (x). We can emphasize this fact by writing ψ E(x) as the solution associated with a particular value of E. But it turns out that it is not all quite as simple as this. To be physically acceptable, the wave function ψ E(x) must satisfy two conditions, one of which we have seen before namely that the wave function must be normalizable, and a second, that the wave function and its derivative must be continuous. Together, these two requirements, the first founded in the probability interpretation of the wave function, the second in more esoteric mathematical necessities which we will not go into here and usually only encountered in somewhat artificial problems, lead to a rather remarkable property of physical systems described by this equation that has enormous physical significance: the quantization of energy.

The Quantization of Energy

At first thought it might seem to be perfectly acceptable to insert any value of E into the time independent Schrödinger equation and solve it for ψ E(x). But in doing so we must remain aware of one further requirement of a wave function which comes from its probability interpretation: to be physically acceptable a wave function must satisfy the normalization condition,

$$\int_{\infty}^{+\infty} \left|\Psi(x,t)\right|^2 dx = 1$$

for all time t. For the particular trial solution introduced above, $\Psi(x,t) = \psi(x)e^{-iEt/\hbar}$:

$$\Psi(x,t) = \psi_E(x)e^{-iEt/\hbar}$$

the requirement that the normalization condition must hold gives, on substituting for Ψ (x,t), the result,

$$\int_{-\infty}^{+\infty} \left|\Psi(x,t)\right|^2 dx = \int_{-\infty}^{+\infty} \left|\psi_E(x)\right|^2 dx = 1$$

Since this integral must be finite, (unity in fact), we must have ψ E(x) → 0 as x → ±∞ in order for the integral to have any hope of converging to a finite value. The importance of this with regard to solving the time dependent Schrödinger equation is that we must check whether or not a solution ψ E(x) obtained for some chosen value of E satisfies the normalization condition. If it does, then this is a physically acceptable solution, if it does not, then that solution and the corresponding value of the energy are not physically acceptable. The particular case of considerable physical significance is if the potential V (x) is attractive, such as would be found with an electron caught by the attractive Coulomb force of an atomic nucleus, or a particle bound by a simple harmonic potential (a mass on a spring), a particle trapped in an infinite potential well. In all such cases, we find that except for certain discrete values of the energy, the wave function ψ E(x) does not vanish, or even worse, diverges, as x → ±∞. In other words, it is only for these discrete values of the energy E that we get physically acceptable wave functions ψ E(x), or to put it more bluntly, the particle can never be observed to have any energy other than these particular values, for which reason these energies are often referred to as the 'allowed' energies of the particle. This pairing off of allowed energy and normalizable wave function is referred to mathematically as ψ E(x) being an eigenfunction of the Schrödinger equation, and E the associated energy eigenvalue, a terminology that acquires more meaning when quantum mechanics is looked at from a more advanced standpoint.

So we have the amazing result that the probability interpretation of the wave function forces us to conclude that the allowed energies of a particle moving in a potential V (x) are restricted to certain discrete values, these values determined by the nature of the potential. This is the phenomenon known as the quantization of energy, a result of quantum mechanics which has enormous significance for determining the structure of atoms, or, to go even further, the properties of matter overall. We have already seen an example of this quantization of energy in our earlier discussion of a particle in an infintely deep potential well, though we did not derive the results by solving the Schrödinger equation itself. We will consider how this is done shortly.

The requirement that Ψ (x) → 0 as x → ±∞ is an example of a boundary condition. Energy quantization is, mathematically speaking, the result of a combined effort: that Ψ (x) be a solution to the time independent Schrödinger equation, and that the solution satisfy these boundary conditions. But both the boundary condition and the Schrödinger equation are derived from, and hence rooted in, the nature of the physical world: we have here an example of the unexpected relevance of purely mathematical ideas in formulating a physical theory.

Continuity Conditions

There is one additional proviso, which was already mentioned briefly above, that has to be applied in some cases. If the potential should be discontinuous in some way, e.g. becoming infinite, as we have seen in the infinite potential well example, or having a finite discontinuity as we will see later in the case of the finite potential well, it is possible for the Schrödinger equation to have solutions that themselves are discontinuous. But discontinuous potentials do not occur in nature (this would imply an infinite force), and as we know that for continuous potentials we always get continuous wave functions, we then place the extra conditions that the wave function and its spatial derivative also must be continuous. We shall see how this extra condition is implemented when we look at the finite potential well later.

Bound States and Scattering States

But what about wave functions such as the harmonic wave function Ψ *(x,t) = A exp[i(kx –* ωt*)*]? These wave functions represent a particle having a definite energy E = $\hbar\omega$ and so would seem to be legitimate and necessary wave functions within the quantum theory. Is that Ψ(x,t) does not vanish as x → ±∞, so the normalization condition, equation $\int_{-\infty}^{+\infty}\left|\Psi(x,t)\right|^2 dx = \int_{-\infty}^{+\infty}\left|\Psi_E(x)\right|^2 dx = 1$ cannot be satisfied. So what is going on here? The answer lies in the fact that there are two kinds of wave functions, those that apply for particles trapped by an attractive potential into what is known as a bound state, and those that apply for particles that are free to travel to infinity (and beyond), otherwise known as scattering states. A particle trapped in an infinitely deep potential well is an example of the former: the particle is confined to move within a restricted region of space. An electron trapped by the attractive potential due to a positively charged atomic nucleus is also an example – the electron rarely moves a distance more than ~10 nm from the nucleus. A nucleon trapped within a nucleus by attractive nuclear forces is yet another. In all these cases, the probability of finding the particle at infinity is zero. In other words, the wave function for the particle satisfies the boundary condition that it vanish at infinity. So we see that it is when a particle is trapped, or confined to a limited region of space by an attractive potential V (*x*) (or V (*r*) in three dimensions), we obtain wave functions that satisfy the above boundary condition, and hand in hand with this, we find that their energies are quantized. But if it should be the case that the particle is free to move as far as it likes in space, in other words, if it is not bound by any attractive potential, (or even repelled by a repulsive potential) then we find that the wave function need not vanish at infinity, and nor is its energy quantized. The problem of how to reconcile this with the normalization condition, and the probability interpretation of the wave function, is a delicate mathematical issue which we cannot hope to address here, but it can be done. Suffice to say that provided the wave function does not diverge at infinity (in other words it remains finite, though not zero) we can give a physical meaning of such states as being an idealized mathematical limiting case which, while it does not satisfy the normalization condition, can still be dealt with in, provided some care is taken with the physical interpretation, in much the same way as the bound state wave functions.

In order to illustrate how the time independent Schrödinger equation can be solved in practice, and some of the characteristics of its solutions, we will here briefly reconsider the infinitely deep potential well problem, already solved by making use of general properties of the wave function. We will then move on to looking at other simple applications.

Solving the Time Independent Schrödinger Equation

The Infinite Potential Well Revisited

Suppose we have a single particle of mass m confined to within a region $0 < x < L$ with potential energy $V = 0$ bounded by infinitely high potential barriers, i.e. $V = \infty$ for $x < 0$ and $x > L$. The potential experienced by the particle is then:

$$\begin{aligned} V(x) &= 0 \qquad 0 < x < L \\ &= \infty \qquad x \geq L; \quad x \leq 0 \end{aligned}$$

In the regions for which the potential is infinite, the wave function will be zero, for exactly the same reasons that it was set to zero, there is zero probability of the particle being found in these regions. Thus, we must impose the boundary conditions:

$$\psi(0) = \psi(L) = 0.$$

Meanwhile, in the region $0 < x < L$, the potential vanishes, so the time independent Schrödinger equation becomes:

$$-\frac{\hbar^2}{2m}\frac{d^2\psi(x)}{dx^2} + E\psi(x).$$

To solve this, we define a quantity k by:

$$k = \sqrt{\frac{2mE}{\hbar^2}}$$

So that $-\frac{\hbar^2}{2m}\frac{d^2\psi(x)}{dx^2} + E\psi(x)$ can be written,

$$\frac{d^2\psi(x)}{dx^2} + k^2\psi(x) = 0$$

Whose general solution is:

$$\psi(x) = A\sin(kx) + B\cos(kx).$$

It is now that we impose the boundary conditions, $\psi(0) = \psi(L) = 0$ to give, first at x = 0:

ψ (0) = B = 0

So that the solution is now:

$$\psi(x) = A\sin(kx).$$

Next, applying the boundary condition at x = L gives,

$$\psi(L) = A\sin(kL) = 0$$

which tells us that either A = 0, in which case ψ (x) = 0, which is not a useful solution (it says that there is no partilce in the well at all!) or else sin(kL) = 0, which gives an equation for k:

$$kL = n\pi, \quad n = 0, \pm1, \pm2, \ \ldots.$$

We exclude the n = 0 possibility as that would give us, once again ψ (x) = 0, and we exclude the negative values of n as the will merely reproduce the same set of solutions (except with opposite sign) as the positive values. Thus we have:

$$k_n = n\pi / L, \ n = 1, 2, \ldots$$

where,we have introduced a subscript n. This leads to, on using equation $k = \sqrt{\frac{2mE}{\hbar^2}}$,

$$E_n = \frac{\hbar^2 k_n^2}{2m} = \frac{n^2\pi^2\hbar^2}{2mL^2}, \ n = 1, 2, \ \ldots$$

Thus we see that the boundary conditions, $\psi(0) = \psi(L) = 0$, have the effect of restricting the values of the energy of the particle to those given by $E_n = \frac{\hbar^2 k_n^2}{2m} = \frac{n^2\pi^2\hbar^2}{2mL^2}, \ n = 1, 2, \ \ldots$. The associated wave functions, that is we apply the normalization condition to determine A (up to an inessential phase factor) which finally gives:

$$\psi_n(x) = \sqrt{\frac{2}{L}}\sin(n\pi x / L) \quad 0 < x < L$$

$$= 0 \qquad\qquad x < 0, x > L.$$

The Finite Potential Well

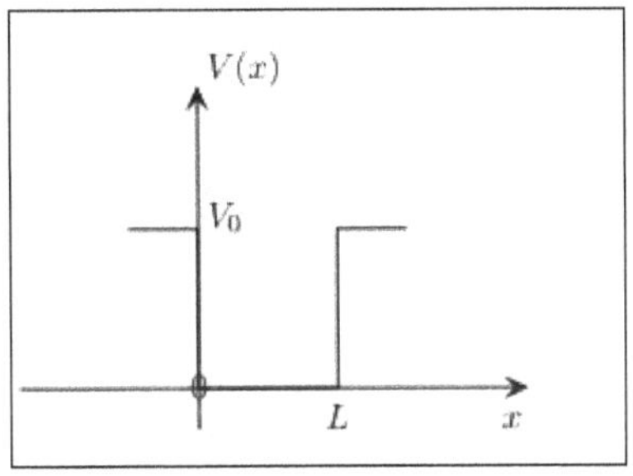

Figure: Finite potential well.

The infinite potential well is a valuable model since, with the minimum amount of fuss, it shows immediately the way that energy quantization as potentials do not occur in nature. However, for electrons trapped in a block of metal, or gas molecules contained in a bottle, this model serves to describe very accurately the quantum character of such systems. In such cases the potential experienced by an electron as it approaches the edges of a block of metal, or as experienced by a gas molecule as it approaches the walls of its container are effectively infinite as far as these particles are concerned, at least if the particles have sufficently low kinetic energy compared to the height of these potential barriers.

But, of course, any potential well is of finite depth, and if a particle in such a well has an energy comparable to the height of the potential barriers that define the well, there is the prospect of the particle escaping from the well. This is true both classically and quantum mechanically, though, as you might expect, the behaviour in the quantum mechanical case is not necessarily consistent with our classical physics based expectations. Thus we now proceed to look at the quantum properties of a particle in a finite potential well.

In this case, the potential will be of the form:

$$\begin{aligned} V(x) &= 0 \quad 0 < x < L \\ &= V \quad x \geq L \quad x \leq 0 \end{aligned}$$

i.e. we have 'lowered' the infinite barriers to a finite value V . We now want to solve the time independent Schrödinger equation for this potential.

To do this, we recognize that the problem can be split up into three parts: x ≤ 0 where the potential is V, 0 < x < L where the potential is zero and x ≥ 0 where the potential is once again V. Therefore, to find the wave function for a particle of energy E, we have to solve three equations, one for each of the regions:

$$\begin{aligned} &\frac{\hbar^2}{2m}\frac{d^2\psi(x)}{dx^2} + (E - V)\psi(x) = 0 \quad x \leq 0 \\ &\frac{\hbar^2}{2m}\frac{d^2\psi(x)}{dx^2} + E\psi = 0 \qquad\qquad 0 < x < L \\ &\frac{\hbar^2}{2m}\frac{d^2\psi(x)}{dx^2} + (E - V)\psi(x) = 0 \quad x \geq L. \end{aligned}$$

The solutions to these equations take different forms depending on whether E < V or E > V. We shall consider the two cases separately.

E < V

First define,

$$k = \sqrt{\frac{2mE}{\hbar^2}} \text{ and } \alpha = \sqrt{\frac{2m(V-E)}{\hbar^2}}$$

Note that, as V > E, α will be a real number, as it is square root of a positive number. We can now write these equations as:

$$\frac{d^2\psi(x)}{dx^2} - \alpha^2\psi(x) = 0 \quad x \le 0$$

$$\frac{d^2\psi(x)}{dx^2} + k^2\psi(x) = 0 \quad 0 < x < L$$

$$\frac{d^2\psi(x)}{dx^2} - \alpha^2\psi(x) = 0 \quad x \ge L$$

Now consider the first of these equations, which will have as its solution:

$$\psi(x) = Ae^{-\alpha x} + Be^{+\alpha x}$$

where A and B are unknown constants. It is at this point that we can make use of our boundary condition, namely that ψ (x) → 0 as x → ±∞. In particular, since the solution we are currently looking at applies for x < 0, we should look at what this solution does for x → −∞. What it does is diverge, because of the term A exp (−αx). So, in order to guarantee that our solution have the correct boundary condition for x → −∞, we must have A = 0. Thus, we conclude that,

$$\psi(x) = Be^{\alpha x} \quad x \le 0$$

We can apply the same kind of argument when solving $\frac{d^2\psi(x)}{dx^2} - \alpha^2\psi(x) = 0$ for x ≥ L. In that case, the solution is:

$$\psi(x) = Ce^{-\alpha x} + De^{\alpha x}$$

but now we want to make certain that this solution goes to zero as x → ∞. To guarantee this, we must have D = 0, so we conclude that:

$$\psi(x) = Ce^{-\alpha x} \quad x \ge L$$

Finally, at least for this part of the argument, we look at the region 0 < x < L. The solution of $\frac{d^2\psi(x)}{dx^2} + k^2\psi(x) = 0$ for this region will be: $\psi(x) = P\cos(kx) + Q\sin(kx) \quad 0 < x < L$.

But now we have no diverging exponentials, so we have to use other means to determine the unknown coefficients P and Q.

At this point we note that we still have four unknown constants B, P, Q, and C. To determine these we note that the three contributions to ψ (x) do not necessarily join together smoothly at x = 0 and x = L. This awkward state of affairs has its origins in the fact that the potential is discontinuous at x = 0 and x = L which meant that we had to solve three separate equations for the three different regions. But these three separate solutions cannot be independent of one another, i.e. there must be a relationship between the unknown constants, so there must be other conditions that enable us to specify these constants. The extra conditions that we impose are that the wave function has to be a continuous function, i.e. the three solutions:

$$\begin{aligned}\psi(x) &= Be^{\alpha x} && x \le 0\\ &= P\cos(kx) + Q\sin(kx) && 0 < x < L\\ &= Ce^{-\alpha x} && x \ge L\end{aligned}$$

should all 'join up' smoothly at x = 0 and x = L. This means that the first two solutions and their slopes (i.e. their first derivatives) must be the same at x = 0, while the second and third solutions and their derivatives must be the same at x = L. Applying this requirement at x = 0 gives:

$$\begin{aligned}&B = P\\ &\alpha B = kQ\end{aligned}$$

and then at x = L:

$$\begin{aligned}&P\cos(kL) + Q\sin(kL) = Ce^{-\alpha L}\\ &-kP\ \sin(kL) + kQ\ \cos(kL) = -\alpha Ce^{-\alpha L}\end{aligned}$$

If we eliminate B and C from these two sets of equations we get, in matrix form:

$$\begin{pmatrix} \alpha & -k \\ \alpha\ \cos(kL) - k\sin(kL) & \alpha\sin(kL)\ +\ k\cos(kL) \end{pmatrix}\begin{pmatrix} P \\ Q \end{pmatrix} = 0$$

and in order that we get a non-trivial solution to this pair of homogeneous equations, the determinant of the coefficients must vanish:

$$\begin{vmatrix} \alpha & -k \\ \alpha\ \cos(kL) - k\sin(kL) & \alpha\sin(kL)\ +\ k\cos(kL) \end{vmatrix} = 0$$

which becomes, after expanding the determinant and rearranging terms:

$$\tan(kL) = \frac{2\alpha k}{k^2 - \alpha^2}.$$

Solving this equation for k will give the allowed values of k for the particle in this finite potential well, and hence, using $k = \sqrt{\frac{2mE}{\hbar^2}}$ and $\alpha = \sqrt{\frac{2m(V-E)}{\hbar^2}}$ in the form:

$$E = \frac{\hbar^2 k^2}{2m}$$

We can determine the allowed values of energy for this particle. What we find is that these allowed energies are finite in number, in contrast to the infinite potential well, but to show this we must solve this equation. This is made difficult to do analytically by the fact that this is a transcendental equation – it has no solutions in terms of familiar functions. However, it is possible to get an idea of what its solutions look like either numerically, or graphically. The latter has some advantages as it allows us to see how the mathematics conspires to produce the quantized energy levels. We can first of all simplify the mathematics a little by writing $\tan(kL) = \frac{2\alpha k}{k^2 - \alpha^2}$ in the form:

$$\tan(kL) = \frac{2(\alpha/k)}{1-(\alpha/k)^2}$$

which, by comparison with the two trigonometric formulae,

$$\tan 2\theta = \frac{2\tan\theta}{1-\tan^2\theta}$$

$$\tan 2\theta = \frac{2\cot(-\theta)}{1-\cot^2(-\theta)}$$

We see that $\tan(kL) = \frac{2\alpha k}{k^2 - \alpha^2}$ is equivalent to the two conditions:

$$\tan\left(\frac{1}{2}kL\right) = \frac{\alpha}{k}$$

$$\cot\left(-\frac{1}{2}kL\right) = -\cot\left(\frac{1}{2}kL\right) = \frac{\alpha}{k}.$$

The aim here is to plot the left and right hand sides of these two expressions as a function of k (well, actually as a function of $-\frac{1}{2}kL$)but before we can do that we need to take account of the fact that the quantity α is given in terms of E by $\sqrt{2m(V-E)/\hbar^2}$, and hence, since $E = \hbar^2 k^2 / 2m$, we have:

$$\frac{\alpha}{k} = \sqrt{\frac{V-E}{E}} = \sqrt{\left(\frac{k_0}{k}\right)^2 - 1}$$

where,

$$k_0 = \sqrt{\frac{2mV}{\hbar^2}}.$$

As we will be plotting as a function of $\frac{1}{2}kL$, it is useful to rewrite the above expression for α/k as:

$$\frac{\alpha}{k} = f\left(\frac{1}{2}kL\right) = \sqrt{\left(\frac{1}{2}k_0 L / \frac{1}{2}kL\right)^2 - 1}.$$

Thus we have,

$$\tan\left(\frac{1}{2}kL\right) = f\left(\frac{1}{2}kL\right) \text{ and } -\cot\left(\frac{1}{2}kL\right) = f\left(\frac{1}{2}kL\right).$$

We can now plot $\tan\left(\frac{1}{2}kL\right), -\cot\left(\frac{1}{2}kL\right)$ and $f\left(\frac{1}{2}kL\right)$ as functions of $\frac{1}{2}kL$ for various values for k_0. The points of intersection of the curve $f\left(\frac{1}{2}kL\right)$ with the tan and cot curves will then give the kL values for an allowed energy level of the particle in this potential.

This is illustrated in figure, where four such plots are given for different values of V. The important feature of these curves is that the number of points of intersection is finite, i.e. there are only a finite number of values of k that solve $\tan(kL) = \frac{2\alpha k}{k^2 - \alpha^2}$. Correspondingly, there will only be a finite number of allowed values of E for the particle, and there will always be at least one allowed value.

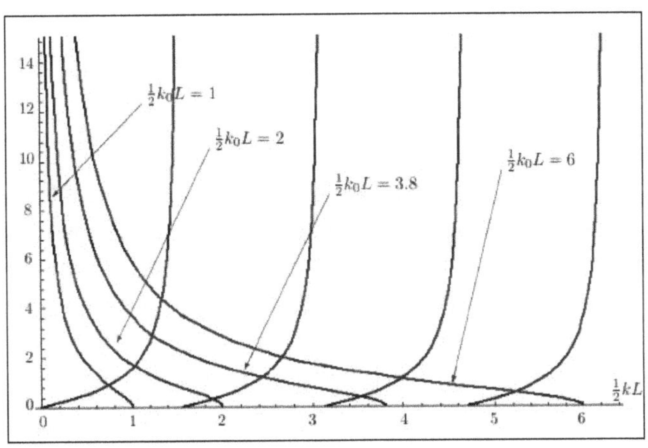

Figure shows, graph to determine bound states of a finite well potential. The points of intersection are solutions to $\tan(kL) = \frac{2\alpha k}{k^2 - \alpha^2}$. The plots are for increasing values of V, starting with V lowest such that $\frac{1}{2}k_0L = 1$, for which there is only one bound state, slightly higher at $\frac{1}{2}k_0L = 2$ for which there are two bound states, slightly higher again for $\frac{1}{2}k_0L = 3.8$ where there are three bound states, and highest of all, $\frac{1}{2}k_0L = 6$ for which there is four bound states.

To determine the corresponding wave functions is a straight forward but complicated task. The first step is to show, by using $\tan(kL) = \frac{2\alpha k}{k^2 - \alpha^2}$ and the equations for B, C, P and Q that:

$$C = e^{\alpha L} B$$

from which readily follows the solution,

$$\begin{aligned}\psi(x) &= Be^{\alpha x} && x \le 0 \\ &= B\left(\cos kx + \frac{\alpha}{k}\sin kx\right) && 0 < x < L \\ &= Be^{-\alpha(x-L)} && x \ge L.\end{aligned}$$

The constant B is determined by the requirement that ψ (x) be normalized, i.e. that,

$$\int_{\infty}^{+\infty} |\psi(x)|^2 \, dx = 1.$$

which becomes:

$$|B|^2\left[\int_{-\infty}^{0} e^{-2\alpha x}\, dx + \int_0^L \left(\cos kx + \frac{\alpha}{k}\sin kx\right)^2 dx + \int_L^{+\infty} e^{-2\alpha(x-L)}\, dx\right] = 1.$$

After a somewhat tedious calculation that makes liberal use of $\tan(kL) = \frac{2\alpha k}{k^2 - \alpha^2}$ the result found is that:

$$B = \frac{k}{k_0}\sqrt{\frac{\alpha}{\frac{1}{2}\alpha L + 1}}.$$

The task of determining the wave functions is then that of determining the allowed values of k from the graphical solution, or numerically, and then substituting those vaules into the above expressions for the wave function. The wave functions found a similar in appearance to the infinte well wave functions, with the big difference that they are non-zero outside the well. This is true even if the particle has the lowest allowed energy, i.e. there is a nonzero probability of finding the particle outside the well. This probability can be readily calculated, being just:

$$P_{outside} = |B|^2\left[\int_{-\infty}^{0} e^{-2\alpha x}\, dx + \int_L^{+\infty} e^{-2\alpha(x-L)}\, dx\right] = \alpha^{-1}|B|^2$$

Scattering from a Potential Barrier

The above examples are of bound states, i.e. wherein the particles are confined to a limited region of space by some kind of attractive or confining potential. However, not all potentials are attractive (e.g. two like charges repel), and in any case, even when there is an attractive potential acting (two opposite charges attracting), it is possible that the particle can be 'free' in the sense that it is not confined to a limited region of space. A simple example of this, classically, is that of a comet orbiting around the sun. It is possible for the comet to follow an orbit in which it initially moves towards the sun, then around the sun, and then heads off into deep space, never to return. This is an example of an unbound orbit, in contrast to the orbits of comets that return repeatedly, though sometimes very infrequently, such as Halley's comet. Of course, the orbiting planets are also in bound states.

A comet behaving in the way just described – coming in from infinity and then ultimately heading off to infinity after bending around the sun – is an example of what is known as a scattering process. In this case, the potential is attractive, so we have the possibility of both scattering occurring, as well as the comet being confined to a closed orbit – a bound state. If the potential was repulsive, then only scattering would occur.

The same distinction applies in quantum mechanics. It is possible for the particle to confined to a limited region in space, in which case the wave function must satisfy the boundary condition that:

$$\psi(x) \to 0 \quad \text{as} \quad x \to \pm\infty$$

As we have seen, this boundary condition is enough to yield the quantization of energy. However, in the quantum analogue of scattering, it turns out that energy is not quantized. This in part can be linked to the fact that the wave function that describes the scattering of a particle of a given energy does not decrease as $x \to \pm\infty$, so that the very thing that leads to the quantization of energy for a bound particle does not apply here.

This raises the question of what to do about the quantization condition, i.e. that:

$$\int_{-\infty}^{+\infty} |\psi(x,t)|^2 \, dx = 1$$

If the wave function does not go to zero as $x \to \pm\infty$, then it is not possible for the wave function to satisfy this normalization condition – the integral will always diverge.

So how are we to maintain the claim that the wave function must have a probability interpretation if one of the principal requirements, the normalization condition, does not hold true? Strictly speaking, a wave function that cannot be normalized to unity is not physically permitted (because it is inconsistent with the probability interpretation of the wave function). Nevertheless, it is possible to retain, and work with, such wave functions, provided a little care is taken. The answer lies in interpreting the wave function so that $|\psi(x,t)|^2 \propto$ particle flux, though we will not be developing this idea to any extent here.

To illustrate the sort of behaviour that we find with particle scattering, we will consider a simple, but important case, which is that of a particle scattered by a potential barrier. This is sufficient to show the sort of things that can happen that agree with our classical intuition, but it also enables us to see that there occurs new kinds of phenomena that have no explanation within classical physics.

Thus, we will investigate the scattering problem of a particle of energy E interacting with a potential V (x) given by:

$$V(x) = 0 \quad x < 0$$
$$V(x) = V_0 \quad x > 0.$$

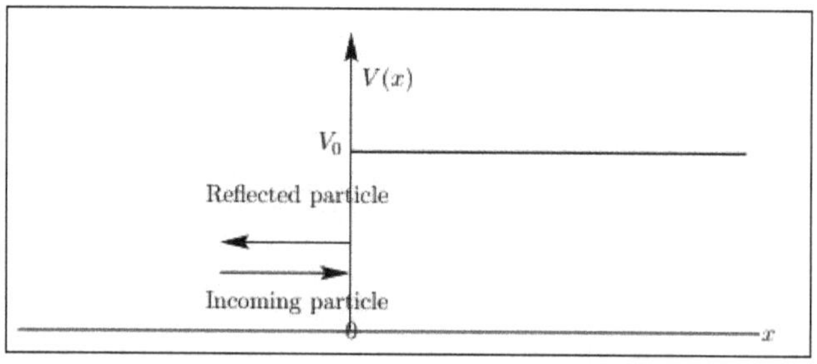

Figure: Potential barrier with particle of energy E < V_0 incident from the left. Classically, the particle will be reflected from the barrier.

In figure above is illustrated what we would expect to happen if a classical particle of energy E < V_0 were incident on the barrier: it would simply bounce back as it has insufficient energy to cross over to x > 0. Quantum mechanically we find that the situation is not so simple.

Given that the potential is as given above, the Schrödinger equation comes in two parts:

$$-\frac{\hbar^2}{2m}\frac{d^2\psi}{dx^2} = E\psi \qquad x<0$$

$$-\frac{\hbar^2}{2m}\frac{d^2\psi}{dx^2} + V_0\psi = E\psi \qquad x>0$$

where E is, once again, the total energy of the particle.

We can rewrite these equations in the following way:

$$\frac{d^2\psi}{dx^2} + \frac{2mE}{\hbar^2}\psi = 0 \quad x<0$$

$$\frac{d^2\psi}{dx^2} - \frac{2mE}{\hbar^2}(V_0 - E)\psi = 0 \quad x>0$$

If we put,

$$k = \frac{\sqrt{2mE}}{\hbar}$$

then the first equation becomes:

$$\frac{d^2\psi}{dx^2} + k^2\psi = 0 \quad x<0$$

which has the general solution:

$$\psi = Ae^{ikx} + Be^{-ikx}$$

where A and B are unknown constants. We can get an idea of what this solution means if we reintroduce the time dependence $\left(with\ \omega = E/\hbar\right)$:

$$\psi(x,t) = \psi(x)e^{-iEt/\hbar} = \psi(x)e^{-i\omega t}$$
$$= Ae^{i(kx-\omega t)} + Be^{-i(kx+\omega t)}$$
$$= \begin{matrix}\text{wave travelling} & + & \text{wave travelling}\\ \text{to right} & & \text{to left}\end{matrix}$$

i.e. this solution represents a wave associated with the particle heading towards the barrier and a reflected wave associated with the particle heading away from the barrier. Later we will see that these two waves have the same amplitude, implying that the particle is perfectly reflected at the barrier.

In the region x > 0, we write,

$$\alpha = \sqrt{2m(V_0 - E)}/\hbar > 0$$

so that the Schrödinger equation becomes,

$$\frac{d^2\psi}{dx^2} - \alpha^2\psi = 0 \quad x > 0$$

which has the solution,

$$\psi = Ce^{-\alpha x} + De^{\alpha x}$$

where C and D are also unknown constants.

The problem here is that the exp(αx) solution grows exponentially with x, and we do not want wave functions that become infinite: it would essentially mean that the particle is forever to be found at x = ∞, which does not make physical sense. So we must put D = 0. Thus, if we put together our two solutions for x < 0 and x > 0, we have:

$$\psi = Ae^{ikx} + Be^{-ikx} \quad x < 0$$
$$= Ce^{-\alpha x} \quad x > 0.$$

If we reintroduce the time dependent factor, we get:

$$\psi(x,t) = \psi(x)e^{-i\omega t} = Ce^{-\alpha x}e^{-i\omega t}$$

which is not a travelling wave at all. It is a stationary wave that simply diminishes in amplitude for increasing x.

We still need to determine the constants A, B, and C. To do this we note that for arbitrary choice of these coefficients, the wave function will be discontinuous at x = 0. For reasons connected with the requirement that probability interpretation of the wave function continue to make physical sense, we will require that the wave function and its first derivative both be continuous at x = 0.

These conditions yield the two equations:

$$C = A + B$$
$$-\alpha C = ik(A - B)$$

which can be solved to give:

$$B = \frac{ik + a}{ik - a}A$$
$$C = \frac{2ik}{ik - \alpha}A$$

and hence,

$$\psi(x) = Ae^{ikx} + \frac{ik + a}{ik - a}Ae^{-ikx} \quad x < 0$$
$$= \frac{2ik}{ik - \alpha}Ae^{-\alpha x} \quad x < 0.$$

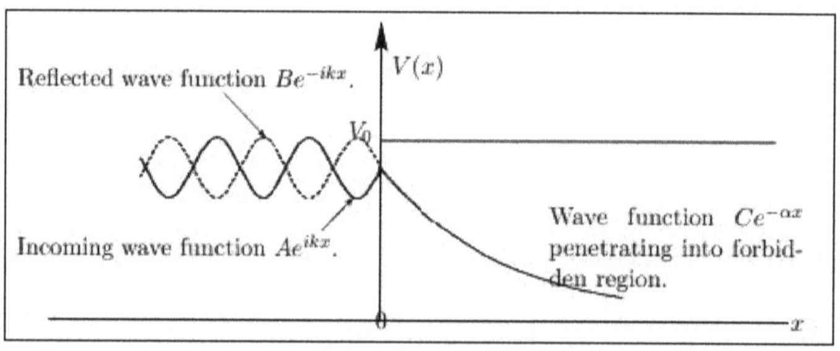

In figure, Potential barrier with wave function of particle of energy E < V_0 incident from the left (solid curve) and reflected wave function (dotted curve) of particle bouncing off barrier. In the clasically 'forbidden' region of x > 0 there is a decaying wave function. Note that the complex wave functions have been represented by their real parts.

Having obtained the mathematical solution, what we need to do is provide a physical interpretation of the result.

First we note that we cannot impose the normalization condition as the wave function does not decrease to zero as x → −∞. But, in keeping with comments made above, we can still learn something from this solution about the behaviour of the particle.

Secondly, we note that the incident and reflected waves have the same 'intensity':

$$\text{Incident intensity} = |A|^2$$

$$\text{Reflected intensity} = |A|^2 \left|\frac{ik + \alpha}{ik - \alpha}\right|^2 = |A|^2$$

and hence they have the same amplitude. This suggests that the incident de Broglie wave is totally reflected, i.e. that the particle merely travels towards the barrier where it 'bounces off', as would be expected classically. However, if we look at the wave function for x > 0 we find that:

$$\begin{aligned}|\psi(x)|^2 &\propto \left|\frac{2ik}{ik-\alpha}\right|^2 e^{-2\alpha x}\\ &= \frac{4k^2}{\alpha^2+k^2}e^{-2\alpha x}\end{aligned}$$

which is an exponentially decreasing probability.

This last result tells us that there is a non-zero probability of finding the particle in the region x > 0 where, classically, the particle has no chance of ever reaching. The distance that the particle can penetrate into this 'forbidden' region is given roughly by 1/2α which, for a subatomic particle can be a few nanometers, while for a macroscopic particle, this distance is immeasurably small.

The way to interpret this result is along the following lines. If we imagine that a particle is fired at the barrier, and we are waiting a short distance on the far side of the barrier in the forbidden region with a 'catcher's mitt' poised to grab the particle then we find that either the particle hits the barrier and bounces off with the same energy as it arrived with, but with the opposite momentum – it never lands in the mitt, or it lands in the mitt and we catch it – it does not bounce off in the opposite direction. The chances of the latter occurring are generally very tiny, but it occurs often enough in microscopic systems that it is a phenomenon that is exploited, particularly in solid state devices. Typically this is done, not with a single barrier, but with a barrier of finite width, in which case the particle can penetrate through the barrier and reappear on the far side, in a process known as quantum tunnelling.

Expectation Value of Momentum

We can make use of Schrödinger's equation to obtain an alternative expression for the expectation value of momentum. This expression is:

$$\langle p\rangle = m\langle v(t)\rangle = m\int_{-\infty}^{+\infty} x\left[\frac{\partial\Psi^*(x,t)}{\partial t}\Psi(x,t)+\Psi^*(x,t)\frac{\partial\Psi(x,t)}{\partial t}\right]dx.$$

We note that the appearance of time derivatives in this expression. If we multiply both sides by $i\hbar$ and make use of Schrödinger's equation, we can substitute for these time derivatives to give:

$$\begin{aligned}i\hbar\langle p\rangle = m\int_{-\infty}^{+\infty} x\Bigg[&\left\{\frac{\hbar^2}{2m}\frac{\partial^2\Psi^*(x,t)}{\partial x^2}-V(x)\Psi^*(x,t)\right\}\Psi(x,t)\\ &+\Psi^*(x,t)\left\{-\frac{\hbar^2}{2m}\frac{\partial^2\Psi(x,t)}{\partial x^2}+V(x)\Psi(x,t)\right\}\Bigg]dx.\end{aligned}$$

The terms involving the potential cancel. The common factor $\hbar^2/2m$ can be moved outside the integral, while both sides of the equation can be divided through by i$\hbar$, yielding a slightly less complicated experssion for $\langle p \rangle$:

$$\langle p \rangle = -\frac{1}{2} i\hbar \int_{-\infty}^{+\infty} x \left[\frac{\partial \Psi^*(x,t)}{\partial x^2} \Psi(x,t) - \Psi^*(x,t) \frac{\partial^2 \Psi(x,t)}{\partial x^2} \right] dx.$$

Integrating both terms in the integrand by parts then gives:

$$\langle p \rangle = \frac{1}{2} i\hbar \int_{-\infty}^{+\infty} \left[\frac{\partial \Psi^*(x,t)}{\partial x} \frac{\partial x \Psi(x,t)}{\partial x} - \frac{\partial x \Psi^*(x,t)}{\partial x} \frac{\partial \Psi(x,t)}{\partial x} \right] dx$$
$$+ \frac{1}{2} i\hbar \left[\frac{\partial \Psi^*(x,t)}{\partial x} \Psi(x,t) - \Psi^*(x,t) \frac{\partial \Psi(x,t)}{\partial x} \right]_{-\infty}^{+\infty}$$

As the wave function vanishes for x → ±∞, the final term here will vanish. Carrying out the derivatives in the integrand then gives:

$$\langle p \rangle = \frac{1}{2} i\hbar \int_{-\infty}^{+\infty} \left[\frac{\partial \Psi^*(x,t)}{\partial x} \Psi(x,t) - \Psi^*(x,t) \frac{\partial \Psi(x,t)}{\partial x} \right] dx$$

Integrating the first term only by parts once again then gives:

$$\langle p \rangle = -i\hbar \int_{-\infty}^{+\infty} \Psi^*(x,t) \frac{\partial \Psi(x,t)}{\partial x} dx + \frac{1}{2} i\hbar \Psi^*(x,t) \Psi(x,t) \Big|_{-\infty}^{+\infty}$$

Once again, the last term here will vanish as the wave function itself vanishes for x → ±∞ and we are left with,

$$\langle p \rangle = -i\hbar \int_{-\infty}^{+\infty} \Psi^*(x,t) \frac{\partial \Psi^*(x,t)}{\partial x} dx.$$

This is a particularly significant result as it shows that the expectation value of momentum can be determined directly from the wave function – i.e. information on the momentum of the particle is contained within the wave function, along with information on the position of the particle. This calculation suggests making the identification:

$$p \rightarrow -i\hbar \frac{\partial}{\partial x}$$

which further suggests that we can make the replacement,

$$p^n \rightarrow \left(-i\hbar \frac{\partial}{\partial x} \right)^n$$

so that, for instance,

$$\langle p^2 \rangle = -\hbar^2 \int_{-\infty}^{+\infty} \Psi^*(x,t) \frac{\partial^2 \Psi(x,t)}{\partial x^2} dx$$

and hence the expectation value of the kinetic energy of the particle is:

$$\langle K\rangle = \frac{\langle p^2\rangle}{2m} = -\frac{\hbar^2}{2m}\int_{-\infty}^{+\infty} \Psi^*(x,t)\frac{\partial^2\Psi(x,t)}{\partial x^2}dx.$$

We can check this idea by turning to the classical formula for the total energy of a particle:

$$\frac{p^2}{2m} + V(x) = E.$$

If we multiply this equation by $\Psi(x,t) = \psi(x)\exp(-iEt/\hbar)$ and make the replacement given in $\langle p^2\rangle = -\hbar^2\int_{-\infty}^{+\infty}\Psi^*(x,t)\frac{\partial^2\Psi(x,t)}{\partial x^2}dx$, we end up with,

$$-\frac{\hbar^2}{2m}\frac{d^2\psi(x)}{dx^2} + V(x)\psi(x) = E\psi(x)$$

which is just the time independent Schrödinger equation. So there is some truth in the ad hoc procedure outlined above.

This association of the physical quantity p with the derivative i.e. is an example of a physical observable, in this case momentum, being represented by a differential operator. This correspondence between physical observables and operators is to be found throughout quantum mechanics. In the simplest case of position, the operator corresponding to position x is itself just x, so there is no subtlties in this case, but as we have just seen this simple state of affairs changes substantially for other observables. Thus, for instance, the observable quantity K, the kinetic energy, is represented by the differential operator:

$$K \to \hat{K} = -\hbar^2\frac{\partial^2}{\partial x^2}.$$

While the operator associated with the position of the particle is $\hat{x}$ with:

$$x \to \hat{x} = x$$

In this last case, the identification is trivial.

Born-Oppenheimer Approximation

The Born-Oppenheimer approximation is one of the basic concepts underlying the description of the quantum states of molecules. This approximation makes it possible to separate the motion of the nuclei and the motion of the electrons. This is not a new idea for us. We already made use of this approximation in the particle-in-a-box model when we explained the electronic absorption spectra of cyanine dyes without considering the motion of the nuclei.

The Born-Oppenheimer approximation neglects the motion of the atomic nuclei when describing the electrons in a molecule. The physical basis for the Born-Oppenheimer approximation is the fact that the mass of an atomic nucleus in a molecule is much larger than the mass of an electron (more than 1000 times). Because of this difference, the nuclei move much more slowly than the electrons. In addition, due to their opposite charges, there is a mutual attractive force of:

$$\frac{Ze^2}{r^2}$$

Acting on an atomic nucleus and an electron. This force causes both particles to be accelerated. Since the magnitude of the acceleration is inversely proportional to the mass, a = f/m, the acceleration of the electrons is large and the acceleration of the atomic nuclei is small; the difference is a factor of more than 1000. Consequently, the electrons are moving and responding to forces very quickly, and the nuclei are not. You can imagine running a 100-yard dash against someone whose acceleration is a 1000 time greater than yours. That person could literally run circles around you. So a good approximation is to describe the electronic states of a molecule by thinking that the nuclei aren't moving, i.e. that they are stationary. The nuclei, however, can be stationary at different positions so the electronic wavefunction can depend on the positions of the nuclei even though their motion is neglected.

Now we look at the mathematics to see what is done in solving the Schrödinger equation after making the Born-Oppenheimer approximation. For a diatomic molecule as an example, the Hamiltonian operator is grouped into three terms:

$$\hat{H}(r,R) = \hat{T}_{nuc}(R) + \frac{e^2}{4\pi\varepsilon_0}\frac{Z_A Z_B}{R} + \hat{H}_{elec}(r,R)$$

where,

$$T_{nuc}(R) = -\frac{\hbar^2}{2m_A}\nabla_A^2 - \frac{\hbar^2}{2m_B}\nabla_B^2$$

and,

$$\hat{H}_{elec}(r,R) = \frac{-\hbar^2}{2m}\sum_i \nabla_i^2 + \frac{e^2}{4\pi\varepsilon_0}\left(-\sum_i \frac{Z_A}{r_{Ai}} - \sum_i \frac{Z_B}{r_{Bi}} + \frac{1}{2}\sum_i \sum_{j\neq i} \frac{1}{r_{ij}}\right)$$

In equation $\hat{H}(r,R) = \hat{T}_{nuc}(R) + \frac{e^2}{4\pi\varepsilon_0}\frac{Z_A Z_B}{R} + \hat{H}_{elec}(r,R)$, the first term represents the kinetic energy of the nuclei, the second term represents the Coulomb repulsion of the two nuclei, and the third term represents the contribution to the energy from the electrons, which consists of their kinetic energy, mutual repulsion for each other, and attraction for the nuclei. Bold-face type is used to represent that r and R are vectors specifying the positions of all the electrons and all the nuclei, respectively.

The Born-Oppenheimer approximation says that the nuclear kinetic energy terms in the complete Hamiltonian, equation $\hat{H}(r,R) = \hat{T}_{nuc}(R) + \frac{e^2}{4\pi\varepsilon_0}\frac{Z_A Z_B}{R} + \hat{H}_{elec}(r,R)$, can be neglected in solving for the electronic wavefunctions and energies. Consequently, the electronic wavefunction φe(r,R) is found as a solution to the electronic Schrödinger equation:

$$\hat{H}_{elec}(r,R)\varphi_e(r,R) = E_e(R)\varphi_e(r,R)$$

Even though the nuclear kinetic energy terms are neglected, the Born-Oppenheimer approximation still takes into account the variation in the positions of the nuclei in determining the electronic energy and the resulting electronic wavefunction depends upon the nuclear positions, R.

As a result of the Born-Oppenheimer approximation, the molecular wavefunction can be written as a product the Born-Oppenheimer Approximation:

$$\psi_{ne}(r,R) = X_{ne}(R)\varphi_e(r,R)$$

This product wavefunction is called the Born-Oppenheimer wavefunction. The function $X_{ne}(R)$ is the vibrational wavefunction, which is a function of the nuclear coordinates RR and depends upon both the vibrational and electronic quantum numbers or states, n and e, respectively. The electronic function, $\varphi_e(r,R)$, is a function of both the nuclear and electronic coordinates, but only depends upon the electronic quantum number or electronic state, e. Translational and rotational motion is not included here. The translational and rotational wavefunctions simply multiply the vibrational and electronic functions in equation $\psi_{ne}(r,R) = X_{ne}(R)\varphi_e(r,R)$ to give he complete molecular wavefunction when the translational and rotational motions are not coupled to the vibrational and electronic motion.

Crude Born-Oppenheimer Approximation

In the *Crude Born-Oppenheimer Approximation*, R is set equal to Ro, the equilibrium separation of the nuclei, and the electronic wavefunctions are taken to be the same for all positions of the nuclei.

The electronic energy, $E_e(R)$, in equation $\hat{H}_{elec}(r,R)\varphi_e(r,R) = E_e(R)\varphi_e(r,R)$ combines with the repulsive Coulomb energy of the two nuclei, to form the potential energy function that controls the nuclear motion.

$$V_e(R) = E_e(R) + \frac{e^2}{4\pi\varepsilon_0}\frac{Z_A Z_B}{R}$$

Consequently the Schrödinger equation for the vibrational motion is:

$$(\hat{T}_{nuc}(R) + V(R))_{X_{ne}}(R) = E_{neX_{ne}}(R)$$

The potential energy was approximated as a harmonic potential depending on the displacement, QQ, of the nuclei from their equilibrium positions.

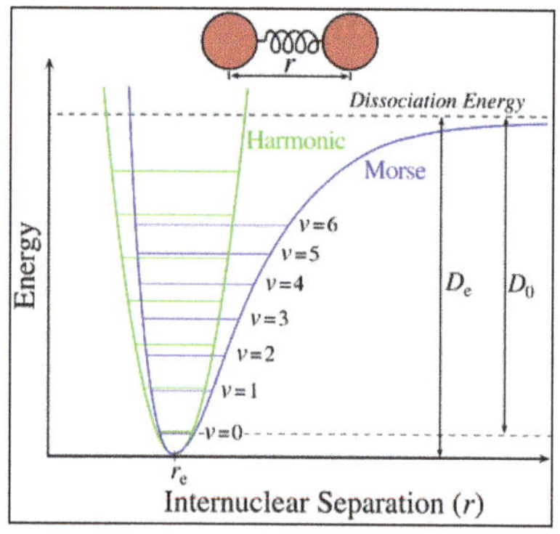

Figure: The potential energy function for a diatomic molecule.

In practice the electronic Schrödinger equation is solved using approximations at particular values of R to obtain the wavefunctions $\varphi_e(r,R)$ and potential energies $V_e(R)$. The potential energies can be graphed as illustrated in figure.

The graph in figure is the energy of a diatomic molecule as a function of internuclear separation, which serves as the potential energy function for the nuclei. When R is very large there are two atoms that are weakly interacting. As R becomes smaller, the interaction becomes stronger, the energy becomes a large negative value, and we say a bond is formed between the atoms. At very small values of R, the internuclear repulsion is very large so the energy is large and positive. This energy function controls the motion of the nuclei. Previously, we approximated this function by a harmonic potential to obtain the description of vibrational motion in terms of the harmonic oscillator model. Other approximate functional forms could be used as well, e.g. the Morse potential. The equilibrium position of the nuclei is where this function is a minimum, i.e. at $R=R_0$. If we obtain the wavefunction at $R=R_0$, and use this function for all values of R, we have employed the *Crude Born-Oppenheimer* approximation.

References

- Quantum-Chemistry: chemistryexplained.com, Retrieved 3 March, 2019
- Lab-science-atom-structure: lwtech.edu, Retrieved 13 July, 2019
- Node: chemistry.gatech.edu, Retrieved 17 January, 2019
- Valence-Bond-Theory, Introduction-to-organic-structure-and-bonding, Organic-Chemistry: libretexts.org, Retrieved 27 March, 2019
- Molecular-orbital-theory, chemical-bonding-and-molecular-structure, chemistry: toppr.com, Retrieved 29 July, 2019
- Density-Functional-Theory, Quantum-Calculations, Quantum-Mechanics, Physical-and-Theoretical-Chemistry: libretexts.org, Retrieved 1 May, 2019
- Chemical-Dynamics, Advanced-Theoretical-Chemistry, Physical-and-Theoretical-Chemistry: libretexts.org, Retrieved 3 August, 2019
- Klopman-Salem-equation: howlingpixel.com, Retrieved 10 April, 2019
- SchrodingerEqn: physics.mq.edu.au, Retrieved 20 February, 2019
- The-Born-Oppenheimer-Approximation, Theories-of-Electronic-Molecular-Structure, Quantum-States-of-Atoms-and-Molecules: libretexts.org, Retrieved 2 June, 2019

5

Equilibrium Chemistry

The state where the properties of a system do not exhibit any observable change due to the rate of forward reaction equaling the rate of backward reaction is known as a chemical equilibrium. Some of the principles which are used to study chemical equilibriums are Le-Chatelier's principle and Nernst distribution law. The chapter closely examines these key concepts of equilibrium chemistry to provide an extensive understanding of the subject.

Chemical Equilibrium

Chemical equilibrium is a state in which the rate of the forward reaction equals the rate of the backward reaction. In other words, there is no net change in concentrations of reactants and products. This kind of equilibrium is also called dynamic equilibrium.

Given the following elementary reaction:

$$aA + bB \underset{k_{-1}}{\overset{k_1}{\rightleftharpoons}} cC + dD$$

The rate of the forward reaction can be represented by the following equation:

$$rate = -k_1[A]^a[B]^b = k_{-1}[C]^c[D]^d$$

where, k_1 and k_{-1} are the reaction constants for the forward and reverse reactions, respectively. Note: the constant k can vary among different reactions at different temperatures.

When the rate is zero, the net concentrations of A, B, C, and D are in equilibrium with each other. Thus, if there is a change in the system due to changes in concentration, temperature, or pressure, the equilibrium will shift to offset the change and re-establish equilibrium according to Le Chatelier's principle. Le Chatelier's principle indicates whether more reactants or more products will be made. Note that the presence of a catalyst does not shift the equilibrium; it only causes the reaction to reach equilibrium faster. This is due to the fact that catalysts only lower activation energies.

Dynamic equilibrium is useful in predicting whether the forward or reverse reaction is spontaneous or nonspontaneous. To explain how this works, three quantities must be introduced: the equilibrium constant, K; the reaction quotient, Q; and the Gibbs Free Energy for a certain reaction, dG_r. The equations for K and Q are given below:

$$K = \frac{a_C^c a_D^d}{a_A^a a_B^b} \text{ at equilibrium}$$

$$Q = \frac{a_C^c a_D^d}{a_A^a a_B^b} \text{ at any point in the reaction}$$

Where, a_C corresponds to the activity of C, a_D corresponds to the activity of D, etc.

The Gibbs free energy of reaction, dG_r, relates to Q and T (temperature) in the following equation:

$$dG_r = dG_r^\circ + RT \ln Q$$

Where, dG_r^o is the Gibbs free energy for the reaction at standard conditions.

When dG_r is negative, the forward reaction is spontaneous and the reverse reaction nonspontaneous. When dG_r is positive, the forward reactions is nonspontaneous and the reverse reaction spontaneous. This is due to Q's deviation from K and Le Chatelier's principle at standard temperature. When Q is larger than K, more products are present initially than at equilibrium, according to Le Chatelier's principle, the equilibrium shifts toward the reactants. In other words, the reverse reaction is spontaneous, which corresponds to a positive dG_r value. When Q is smaller than K, more reactants are present initially than at equilibrium, the reaction shifts towards the products, thus the forward reaction is spontaneous. This corresponds to a negative dG_r value. The relationship between Q and dG_r can be represented by the following graph:

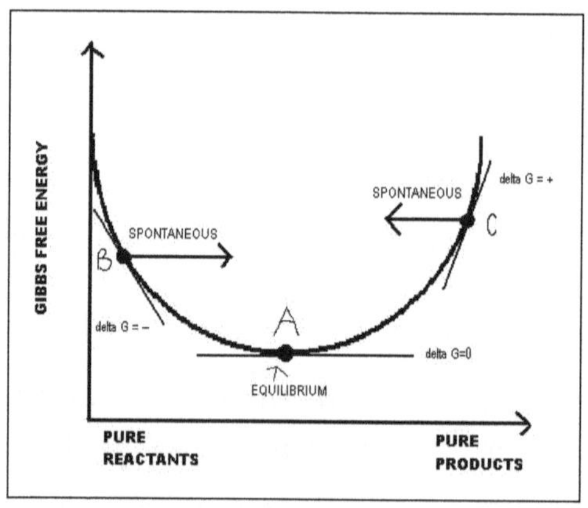

At point A, the system is in equilibrium; therefore, Q=K and dG_r=0. At point B, Q<K; the forward reaction is therefore spontaneous in order to reach equilibrium, and dG_r<0. At point C, Q>K, and thus the reverse reaction is spontaneous in order to reach equilibrium; dG_r>0.

- An equilibrium constant expression includes terms only for reactants and products whose concentrations and/or partial pressures change during a chemical reaction.
- If an equilibrium system is subjected to a change in temperature, pressure, or concentration of a reacting species, the system will respond with a new equilibrium state.
- When the volume of an equilibrium mix of gases is reduced, a net change will occur in the direction that produces fewer moles of gas.
- When the volume of an equilibrium mix of gases is increased, a net change will occur in the direction that produces more moles of gas.

Equilibrium Constants

The equilibrium constant of a chemical reaction is the value of its reaction quotient at chemical equilibrium.

For a general chemical reaction occurring in solution,

$$aA + bB \rightleftarrows cC + dD$$

the equilibrium constant, also known as K_{eq}, is defined by the following expression:

$$K_{eq} = \frac{[C]^c[D]^d}{[A]^a[B]^b}$$

where, [A] is the molar concentration of species A at equilibrium, and so forth. The coefficients a, b, c, and d in the chemical equation become exponents in the expression for K_{eq}. The K_{eq} is a characteristic numerical value for a given reaction at a given temperature; that is, each chemical reaction has its own characteristic K_{eq}. The concentration of each reactant and product in a chemical reaction at equilibrium is *related*; the concentrations cannot be random values, but they depend on each other. The numerator of the expression for K_{eq} has the concentrations of every product (however many products there are), while the denominator of the expression for K_{eq} has the concentrations of every reactant, leading to the common *products over reactants* definition for the K_{eq}.

Let us consider a simple example. Suppose we have this equilibrium:

$$A \rightleftarrows B$$

There is one reactant, one product, and the coefficients on each are just 1 (assumed, not written). The K_{eq} expression for this equilibrium is:

$$K_{eq} = \frac{[B]}{[A]}$$

(Exponents of 1 on each concentration are understood.) Suppose the numerical value of K_{eq} for this chemical reaction is 2.0. If [B] = 4.0 M, then [A] must equal 2.0 M so that the value of the fraction equals 2.0:

$$K_{eq} = \frac{[B]}{[A]} = \frac{4.0}{2.0} = 2.0$$

By convention, the units are understood to be M and are omitted from the K_{eq} expression. Suppose [B] were 6.0 M. For the K_{eq} value to remain constant (it is, after all, called the equilibrium *constant*), then [A] would have to be 3.0 M at equilibrium:

$$K_{eq} = \frac{[B]}{[A]} = \frac{6.0}{3.0} = 2.0$$

If [A] were *not* equal to 3.0 M, the reaction would not be at equilibrium, and a net reaction would occur until that ratio was indeed 2.0. At that point, the reaction is at equilibrium, and any net change would cease.

The issue is the same with more complex expressions for the K_{eq}; only the mathematics becomes more complex. Generally speaking, given a value for the K_{eq} and all but one concentration at equilibrium, the missing concentration can be calculated.

In some types of equilibrium problems, square roots, cube roots, or even higher roots need to be analyzed to determine a final answer. Make sure you know how to perform such operations on your calculator; if you do not know, ask your instructor for assistance.

The K_{eq} was defined earlier in terms of concentrations. For gas-phase reactions, the K_{eq} can also be defined in terms of the partial pressures of the reactants and products, P_i. For the gas-phase reaction:

$$aA(g) + bB(g) \rightleftarrows cC(g) + dD(g)$$

The pressure-based equilibrium constant, K_P, is defined as follows:

$$K_P = \frac{P_C^c P_D^d}{P_A^a P_B^b}$$

where, P_A is the partial pressure of substance A at equilibrium in atmospheres, and so forth. As with the concentration-based equilibrium constant, the units are omitted when substituting into the expression for K_P.

There is a simple relationship between K_{eq} (based on concentration units) and K_P (based on pressure units):

$$K_P = K_{eq}.(RT)^{\Delta n}$$

where, R is the ideal gas law constant (in units of L·atm/mol·K), T is the absolute temperature, and Δn is the change in the number of moles of gas in the balanced chemical equation, defined as $n_{gas,prods} - n_{gas,rcts}$. This equation implies that if the number of moles of gas are the same in reactants and products, $K_{eq} = K_P$.

Finally, we recognize that many chemical reactions involve substances in the solid or liquid phases. For example, a particular chemical reaction is represented as follows:

$$2\ NaHCO_3(s) \rightleftarrows Na_2CO_3(s) + CO_2(g) + H_2O(\ell)$$

This chemical equation includes all three phases of matter. This kind of equilibrium is called a heterogeneous equilibrium because there is more than one phase present.

The rule for heterogeneous equilibria is as follows: *Do not include the concentrations of pure solids and pure liquids in* K_{eq} *expressions.* Only partial pressures for gas-phase substances or concentrations in solutions are included in the expressions of equilibrium constants. As such, the equilibrium constant expression for this reaction would simply be:

$$KP = PCO_2$$

because the two solids and one liquid would not appear in the expression.

Gibbs Free Energy

Some reactions are spontaneous because they give off energy in the form of heat ($\Delta H < 0$). Others are spontaneous because they lead to an increase in the disorder of the system ($\Delta S > 0$). Calculations of ΔH and ΔS can be used to probe the driving force behind a particular reaction.

When one of the potential driving forces behind a chemical reaction is favorable and the other is not, a new quantity is introduced known as the Gibbs free energy (G) of the system which reflects the balance between these forces.

The Gibbs free energy of a system at any moment in time is defined as the enthalpy of the system minus the product of the temperature times the entropy of the system.

$$G = H - TS$$

The Gibbs free energy of the system is a state function because it is defined in terms of thermodynamic properties that are state functions. The change in the Gibbs free energy of the system that occurs during a reaction is therefore equal to the change in the enthalpy of the system minus the change in the product of the temperature times the entropy of the system.

$$\Delta G = \Delta H - \Delta(TS)$$

If the reaction is run at constant temperature, this equation can be written as follows:

$$\Delta G = \Delta H - T\Delta S$$

The change in the free energy of a system that occurs during a reaction can be measured under any set of conditions. If the data are collected under standard-state conditions, the result is the standard-state free energy of reaction (ΔG^o).

$$\Delta G^o = \Delta H^o - T\Delta S^o$$

The beauty of the equation defining the free energy of a system is its ability to determine the relative importance of the enthalpy and entropy terms as driving forces behind a particular reaction. The change in the free energy of the system that occurs during a reaction measures the balance between the two driving forces that determine whether a reaction is spontaneous. As we have seen, the enthalpy and entropy terms have different sign conventions.

Favorable		*Unfavorable*
$\Delta H^o < 0$		$\Delta H^o > 0$
$\Delta S^o > 0$		$\Delta S^o < 0$

The entropy term is therefore subtracted from the enthalpy term when calculating ΔG^o for a reaction.

Because of the way the free energy of the system is defined, ΔG^o is negative for any reaction for which ΔH^o is negative and ΔS^o is positive. ΔG^o is therefore negative for any reaction that is favored by both the enthalpy and entropy terms. We can therefore conclude that any reaction for which ΔG^o is negative should be favorable, or spontaneous.

Favorable, or spontaneous reactions: $\Delta G^o < 0$

Conversely, ΔG^o is positive for any reaction for which ΔH^o is positive and ΔS^o is negative. Any reaction for which ΔG^o is positive is therefore unfavorable.

Unfavorable, or non-spontaneous reactions: $\Delta G^o > 0$

Reactions are classified as either exothermic ($\Delta H < 0$) or endothermic ($\Delta H > 0$) on the basis of whether they give off or absorb heat. Reactions can also be classified as exergonic ($\Delta G < 0$) or endergonic ($\Delta G > 0$) on the basis of whether the free energy of the system decreases or increases during the reaction.

When a reaction is favored by both enthalpy ($\Delta H^o < 0$) and entropy ($\Delta S^o > 0$), there is no need to calculate the value of ΔG^o to decide whether the reaction should proceed. The same can be said for reactions favored by neither enthalpy ($\Delta H^o > 0$) nor entropy ($\Delta S^o < 0$). Free energy calculations become important for reactions favored by only one of these factors.

The Effect of Temperature on the Free Energy of a Reaction

The balance between the contributions from the enthalpy and entropy terms to the free energy of a reaction depends on the temperature at which the reaction is run.

The equation used to define free energy suggests that the entropy term will become more important as the temperature increases.

$$\Delta G^o = \Delta H^o - T\Delta S^o$$

Since the entropy term is unfavorable, the reaction should become less favorable as the temperature increases.

Standard-State Free Energies of Reaction

ΔG^o for a reaction can be calculated from tabulated standard-state free energy data. Since there is no absolute zero on the free-energy scale, the easiest way to tabulate such data is in terms of standard-state free energies of formation, ΔG_f^o. As might be expected, the standard-state free energy of formation of a substance is the difference between the free energy of the substance and the free energies of its elements in their thermodynamically most stable states at 1 atm, all measurements being made under standard-state conditions.

Interpreting Standard-State Free Energy of Reaction Data

What does the value of ΔG^o tell us about the following reaction:

$$N_2(g) + 3\,H_2(g) \rightleftharpoons 2\,NH_3(g)$$

$$\Delta G^o = -32.96 \text{ kJ}$$

By definition, the value of ΔG^o for a reaction measures the difference between the free energies of the reactants and products when all components of the reaction are present at standard-state conditions.

ΔG^o therefore describes this reaction only when all three components are present at 1 atm pressure.

The *sign* of ΔG^o tells us the direction in which the reaction has to shift to come to equilibrium. The fact that ΔG^o is negative for this reaction at 25 °C means that a system under standard-state conditions at this temperature would have to shift to the right, converting some of the reactants into products, before it can reach equilibrium. The *magnitude* of ΔG^o for a reaction tells us how far the standard state is from equilibrium. The larger the value of ΔG^o, the further the reaction has to go to get to from the standard-state conditions to equilibrium.

Assume, for example, that we start with the following reaction under standard-state conditions, as shown in the figure.

$$N_2(g) + 3\,H_2(g) \rightleftharpoons 2\,NH_3(g)$$

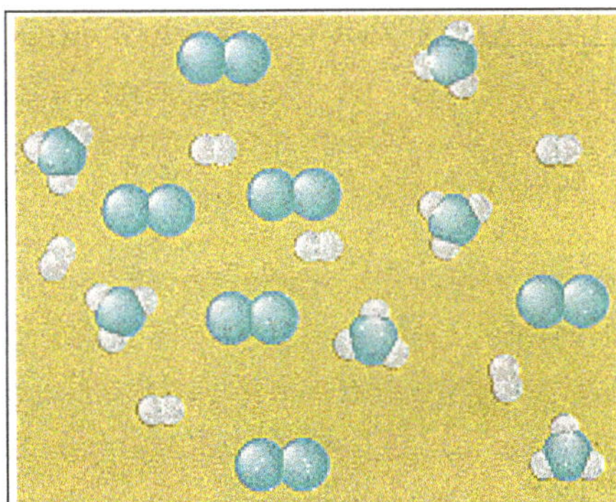

Standard State

$P_{H_2} = P_{N_2} = P_{NH_3} = 1$ atm

The value of ΔG at that moment in time will be equal to the standard-state free energy for this reaction, ΔG^o.

When, $Q_p = 1$:

$$\Delta G = \Delta G^o$$

As the reaction gradually shifts to the right, converting N_2 and H_2 into NH_3, the value of ΔG for the reaction will decrease. If we could find some way to harness the tendency of this reaction to come to equilibrium, we could get the reaction to do work. The free energy of a reaction at any moment in time is therefore said to be a measure of the energy available to do work.

The Relationship between Free Energy and Equilibrium Constants

When a reaction leaves the standard state because of a change in the ratio of the concentrations of the products to the reactants, we have to describe the system in terms of non-standard-state free energies of reaction. The difference between ΔG^o and ΔG for a reaction is important. There is only one value of ΔG^o for a reaction at a given temperature, but there are an infinite number of possible values of ΔG.

The figure shows the relationship between ΔG for the following reaction and the logarithm to the base *e* of the reaction quotient for the reaction between N_2 and H_2 to form NH_3.

$$N_2(g) + 3\,H_2(g) \rightleftharpoons 2\,NH_3(g)$$

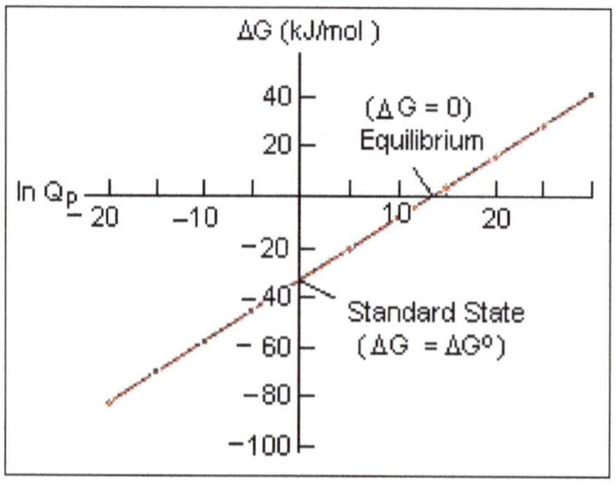

Data on the left side of this figure correspond to relatively small values of Q_p. They therefore describe systems in which there is far more reactant than product. The sign of ΔG for these systems is negative and the magnitude of ΔG is large. The system is therefore relatively far from equilibrium and the reaction must shift to the right to reach equilibrium.

Data on the far right side of this figure describe systems in which there is more product than reactant. The sign of ΔG is now positive and the magnitude of ΔG is moderately large. The sign of ΔG tells us that the reaction would have to shift to the left to reach equilibrium. The magnitude of ΔG tells us that we don't have quite as far to go to reach equilibrium.

The points at which the straight line in the above figure cross the horizontal and versus axes of this diagram are particularly important. The straight line crosses the vertical axis when the reaction quotient for the system is equal to 1. This point therefore describes the standard-state conditions, and the value of ΔG at this point is equal to the standard-state free energy of reaction, ΔG^o.

When $Q_p = 1$:

$$\Delta G = \Delta G^o$$

The point at which the straight line crosses the horizontal axis describes a system for which ΔG is equal to zero. Because there is no driving force behind the reaction, the system must be at equilibrium.

When $Q_p = K_p$:

$$\Delta G = 0$$

The relationship between the free energy of reaction at any moment in time (ΔG) and the standard-state free energy of reaction (ΔG^o) is described by the following equation:

$$\Delta G = \Delta G^o + RT \ln Q$$

In this equation, *R* is the ideal gas constant in units of J/mol-K, *T* is the temperature in kelvin, *ln* represents a logarithm to the base *e*, and *Q* is the reaction quotient at that moment in time.

As we have seen, the driving force behind a chemical reaction is zero ($\Delta G = 0$) when the reaction is at equilibrium ($Q = K$).

$$0 = \Delta G^{o} + RT \ln K$$

We can therefore solve this equation for the relationship between ΔG^{o} and K.

$$\Delta G^{o} = - RT \ln K$$

This equation allows us to calculate the equilibrium constant for any reaction from the standard-state free energy of reaction, or vice versa.

The key to understanding the relationship between G^{o} and K is recognizing that the magnitude of ΔG^{o} tells us how far the standard-state is from equilibrium. The smaller the value of ΔG^{o}, the closer the standard-state is to equilibrium. The larger the value of ΔG^{o}, the further the reaction has to go to reach equilibrium. The relationship between ΔG^{o} and the equilibrium constant for a chemical reaction is illustrated by the data in the table.

Table: Values of ΔG^{o} and K for Common Reactions at 25 °C.

Reaction	ΔG^{o} *(kJ)*	*K*
$2\ SO_3(g) \rightleftharpoons 2\ SO_2(g) + O_2(g)$	141.7	1.4×10^{-25}
$H_2O(l) \rightleftharpoons H^+(aq) + OH^-(aq)$	79.9	1.0×10^{-14}
$AgCl(s) + H_2O \rightleftharpoons Ag^+(aq) + Cl^-(aq)$	55.6	1.8×10^{-10}
$HOAc(aq) + H_2O \rightleftharpoons H^+(aq) + OAc^-(aq)$	27.1	1.8×10^{-5}
$N_2(g) + 3\ H_2(g) \rightleftharpoons 2\ NH_3(g)$	-32.9	5.8×10^{5}
$HCl(aq) + H_2O \rightleftharpoons H^+(aq) + Cl^-(aq)$	-34.2	1×10^{6}
$Cu^{2+}(aq) + 4\ NH_3(aq) \rightleftharpoons Cu(NH_3)_4^{2+}(aq)$	-76.0	2.1×10^{13}
$Zn(s) + Cu^{2+}(aq) \rightleftharpoons Zn^{2+}(aq) + Cu(s)$	-211.8	1.4×10^{37}

The equilibrium constant for a reaction can be expressed in two ways: K_c and K_p. We can write equilibrium constant expressions in terms of the partial pressures of the reactants and products, or in terms of their concentrations in units of moles per liter.

For gas-phase reactions the equilibrium constant obtained from ΔG^{o} is based on the partial pressures of the gases (K_p). For reactions in solution, the equilibrium constant that comes from the calculation is based on concentrations (K_c).

The Temperature Dependence of Equilibrium Constants

Equilibrium constants are not strictly constant because they change with temperature. We are now ready to understand why.

The standard-state free energy of reaction is a measure of how far the standard-state is from equilibrium.

$$\Delta G^{o} = - RT \ln K$$

But the magnitude of ΔG^{o} depends on the temperature of the reaction.

$$\Delta G^{o} = \quad H^{o} - T \quad S^{o}$$

As a result, the equilibrium constant must depend on the temperature of the reaction.

A good example of this phenomenon is the reaction in which NO_2 dimerizes to form N_2O_4.

$$2\ NO_2(g) \rightleftharpoons N_2O_4(g)$$

This reaction is favored by enthalpy because it forms a new bond, which makes the system more stable. The reaction is not favored by entropy because it leads to a decrease in the disorder of the system.

NO_2 is a brown gas and N_2O_4 is colorless. We can therefore monitor the extent to which NO_2 dimerizes to form N_2O_4 by examining the intensity of the brown color in a sealed tube of this gas. What should happen to the equilibrium between NO_2 and N_2O_4 as the temperature is lowered?

For the sake of argument, let's assume that there is no significant change in either ΔH^{o} or ΔS^{o} as the system is cooled. The contribution to the free energy of the reaction from the enthalpy term is therefore constant, but the contribution from the entropy term becomes smaller as the temperature is lowered.

$$\Delta G^{o} = \Delta H^{o} - T\Delta S^{o}$$

As the tube is cooled, and the entropy term becomes less important, the net effect is a shift in the equilibrium toward the right. The figure below shows what happens to the intensity of the brown color when a sealed tube containing NO_2 gas is immersed in liquid nitrogen. There is a drastic decrease in the amount of NO_2 in the tube as it is cooled to -196 °C.

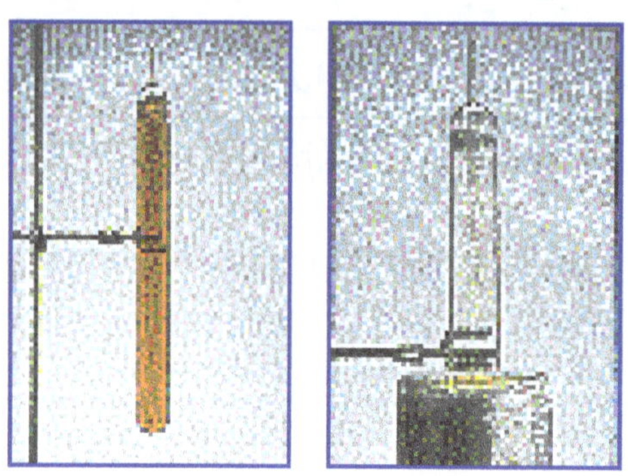

The Relationship between Free Energy and Cell Potentials

The value of G for a reaction at any moment in time tells us two things. The sign of ΔG tells us in what direction the reaction has to shift to reach equilibrium. The magnitude of ΔG tells us how far the reaction is from equilibrium at that moment.

The potential of an electrochemical cell is a measure of how far an oxidation-reduction reaction is from equilibrium. The Nernst equation describes the relationship between the cell potential at any moment in time and the standard-state cell potential.

$$E = E^{O} - \frac{RT}{nF}\ln Q_c$$

Let's rearrange this equation as follows:

$$nFE = nFE^{o} - RT\ln Q$$

We can now compare it with the equation used to describe the relationship between the free energy of reaction at any moment in time and the standard-state free energy of reaction.

$$\Delta G = \Delta G^{o} + RT\ln Q$$

These equations are similar because the Nernst equation is a special case of the more general free energy relationship. We can convert one of these equations to the other by taking advantage of the following relationships between the free energy of a reaction and the cell potential of the reaction when it is run as an electrochemical cell.

$$\Delta G = -nFE$$

Le Chatelier's Principle

Le Chatelier's Principle is the principle when a stress is applied to a chemical system at equilibrium, the equilibrium will shift to relieve the stress. In other words, it can be used to predict the direction of a chemical reaction in response to a change in conditions of temperature, concentration, volume, or pressure. While Le Chatelier's principle can be used to predict the response to a change in equilibrium, it does not explain (at a molecular level), why the system responds as it does.

Chatalier's Principle or the Equilibrium Law

The principle is named for Henry Louis Le Chatelier. Le Chatelier and Karl Ferdinand Braun independently proposed the principle, which is also known as Chatelier's principle or the equilibrium law. The law may be stated:

> "When a system at equilibrium is subjected to a change in temperature, volume, concentration, or pressure, the system readjusts to partially counter the effect of the change, resulting in a new equilibrium."

While chemical equations are typically written with reactants on the left, an arrow pointing from left to right, and products on the right, the reality is that a chemical reaction is at equilibrium. In other words, a reaction may proceed in both the forward and backward direction or be reversible. At equilibrium, both the forward and back reactions occur. One may proceed much more quickly than the other.

In addition to chemistry, the principle also applies, in slightly different forms, to the fields of pharmacology and economics.

Use of Le Chatelier's Principle in Chemistry

1. Concentration: An increase in the amount of reactants (their concentration) will shift the equilibrium to produce more products (product-favored). Increasing the number of products will shift the reaction to make more reactants (reactant-favored). Decreasing reactants favors reactants. Decreasing product favors products.

2. Temperature: Temperature may be added to a system either externally or as a result of the chemical reaction. If a chemical reaction is exothermic (ΔH is negative or heat is released), heat is considered a product of the reaction. If the reaction is endothermic (ΔH is positive or heat is absorbed), heat is considered a reactant. So, increasing or decreasing temperature can be considered the same as increasing or decreasing the concentration of reactants or products. In the temperature is increased, the heat of the system increases, causing the equilibrium to shift to the left (reactants). If the temperature is decreased, the equilibrium shifts to the right (products). In other words, the system compensates for the reduction in temperature by favoring the reaction that generates heat.

3. Pressure/Volume: Pressure and volume can change if one or more of the participants in a chemical reaction is a gas. Changing the partial pressure or volume of a gas acts the same as changing its concentration. If the volume of gas increases, pressure decreases (and vice versa). If the pressure or volume increases, the reaction shifts toward the side with lower pressure. If the pressure is increased or volume decreases, equilibrium shifts toward the higher pressure side of the equation. Note, however, that adding an inert gas (e.g., argon or neon) increases the overall pressure of the system, yet does not change the partial pressure of the reactants or products, so no equilibrium shift occurs.

Debye–Hückel Theory

The theory of Peter Debye and Erich Hückel provides theoretical expressions for single-ion activity coefficients and mean ionic activity coefficients in electrolyte solutions. The expressions in one form or another are very useful for extrapolation of quantities that include mean ionic activity coefficients to low solute molality or infinite dilution.

The only interactions the theory considers are the electrostatic interactions between ions. These interactions are much stronger than those between uncharged molecules, and they die off more slowly with distance. If the positions of ions in an electrolyte solution were completely random, the net effect of electrostatic ion–ion interactions would be zero, because each cation–cation or anion–anion repulsion would be balanced by a cation–anion attraction. The positions are not random, however: each cation has a surplus of anions in its immediate environment, and each anion has a surplus of neighboring cations. Each ion therefore has a net attractive interaction with the surrounding ion atmosphere. The result for a cation species at low electrolyte molality is a decrease of $\mu+$ compared to the cation at same molality in the absence of ion–ion interactions, meaning that the single-ion activity coefficient γ+ becomes less than 1 as the electrolyte molality is increased beyond the ideal-dilute range. Similarly, γ− also becomes less than 1.

According to the Debye–Hückel theory, the single-ion activity coefficient γi of ion i in a solution of one or more electrolytes is given by:

$$\ln \gamma_i = -\frac{A_{DH} z_i^2 \sqrt{I_m}}{1 + B_{DH} a \sqrt{I_m}}$$

where,

- z_i= the charge number of ion i (+1, −2, etc.);
- I_m= the ionic strength of the solution on a molality basis, defined by:

$$I_m \stackrel{\text{def}}{=} \frac{1}{2} \sum_{\text{all ions}} m_j z_j^2$$

- A_{DH} and B_{DH} are defined functions of the kind of solvent and the temperature;
- a is an adjustable parameter, equal to the mean effective distance of closest approach of other ions in the solution to one of the i ions.

The definitions of the quantities A_{DH} and B_{DH} appearing in $\ln \gamma_i = -\frac{A_{DH} z_i^2 \sqrt{I_m}}{1 + B_{DH} a \sqrt{I_m}}$ are:

$$A_{\text{DH}} \stackrel{\text{def}}{=} \left(N_A^2 e^3/8\pi\right)\left(2\rho_A^*\right)^{1/2}\left(\epsilon_r \epsilon_0 RT\right)^{-3/2}$$

$$B_{\text{DH}} \stackrel{\text{def}}{=} N_A e\left(2\rho_A^*\right)^{1/2}\left(\epsilon_r \epsilon_0 RT\right)^{-1/2}$$

where, N_A is the Avogadro constant, e is the elementary charge (the charge of a proton), ρ^*_A and ϵ_r are the density and relative permittivity (dielectric constant) of the solvent, and ϵ_0 is the electric constant (or permittivity of vacuum).

Lewis and Randall introduced the term *ionic strength*, defined by $I_m \stackrel{\text{def}}{=} \frac{1}{2} \sum_{\text{all ions}} m_j z_j^2$, two years before the Debye–Hückel theory was published. They found empirically that in dilute solutions, the mean ionic activity coefficient of a given strong electrolyte is the same in all solutions having the same ionic strength.

When the solvent is water at 25 °C, the quantities A_{DH} and B_{DH} have the values:

$$A_{\text{DH}} = 1.1744\,\text{kg}^{1/2}\ \text{mol}^{-1/2}$$

$$B_{\text{DH}} = 3.285 \times 10^9\ \text{m}^{-1}\ \text{kg}^{1/2}\ \text{mol}^{-1/2}$$

From $\ln \gamma_i = -\frac{A_{DH} z_i^2 \sqrt{I_m}}{1 + B_{DH} a \sqrt{I_m}}$ and $\ln \gamma_i = -\frac{A_{DH} z_i^2 \sqrt{I_m}}{1 + B_{DH} a \sqrt{I_m}}$ the electroneutrality condition $v_+ z_+ = v_- z_-$, we obtain the following expression for the logarithm of the mean ionic activity coefficient of an electrolyte solute:

$$\ln \gamma \pm = -\frac{A_{DH}\left|z_{+}z_{-}\right|\sqrt{I_m}}{1 + B_{DH}a\sqrt{I_m}}$$

In this equation, z_+ and z_- are the charge numbers of the cation and anion of the solute. Since the right side of $\ln \gamma \pm = -\frac{A_{DH}\left|z + z -\right|\sqrt{I_m}}{1 + B_{DH}a\sqrt{I_m}}$ is negative at finite solute molalities, and zero at infinite dilution, the theory predicts that γ± is less than 1 at finite solute molalities and approaches 1 at infinite dilution.

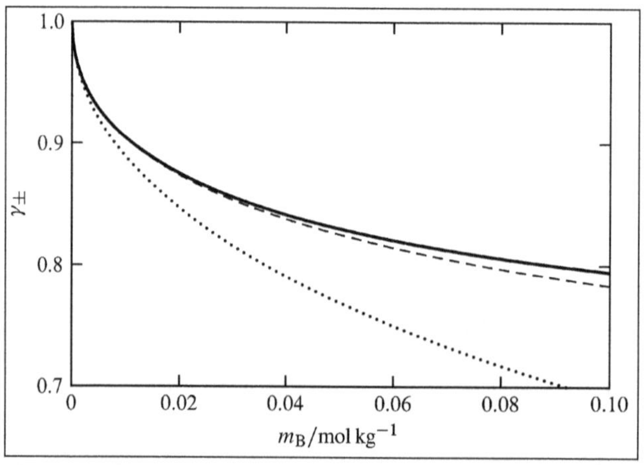

Figure: Mean ionic activity coefficient of aqueous HCl at 25 °C. Solid curve: experiment; dashed curve: Debye–Hückel theory with a=5×10⁻¹⁰ma; dotted curve: Debye–Hückel limiting law.

Figure shows that with the proper choice of the parameter aa, the mean ionic activity coefficient of HCl calculated from $\ln \gamma \pm = -\frac{A_{DH}\left|z + z -\right|\sqrt{I_m}}{1 + B_{DH}a\sqrt{I_m}}$ (dashed curve) agrees closely with experiment (solid curve) at low molalities.

As the molalities of all solutes become small, $\ln \gamma \pm = -\frac{A_{DH}\left|z + z -\right|\sqrt{I_m}}{1 + B_{DH}a\sqrt{I_m}}$ becomes:

$$\ln \gamma \pm = -A_{DH}\left|z_{+}z_{-}\right|\sqrt{I_m}$$

This form is known as the Debye–Hückel limiting law. Note that the limiting law contains no adjustable parameters. The dotted curve in figure shows that the limiting law agrees with experiment only at quite low molality.

The ionic strength I_m is calculated from $I_m \overset{\text{def}}{=} \frac{1}{2}\sum_{\text{all ions}} m_j z_j^2$ with the molalities of *all* ions in the solution, not just the molality of the ion or solute whose activity coefficient we are interested in. This is because, as explained above, the departure of γ_+ and γ_- from the ideal-dilute value of 1 is caused by the interaction of each ion with the ion atmosphere resulting from all other ions in the solution.

In a binary solution of a single electrolyte solute, assumed to be completely dissociated, the relation between the ionic strength and the solute molality depends on ν (the number of ions per solute formula unit) and the charge numbers z_+ and z_-. The ionic strength is given by:

$$I_m = (1/2)\sum_i m_i z_i^2 = (1/2)(\nu_+ z_+^2 + \nu_- z_-^2)m_B.$$

With the help of the electroneutrality condition $\nu_{+z+} = -(\nu_{-z-})$, this becomes:

$$\begin{aligned} I_m &= \frac{1}{2}[-(\nu_- z_-)z_+ - (\nu_+ z_+)z_-]m_B \\ &= \frac{1}{2}[-(\nu_- + \nu_+)z_+ z_-]m_B \\ &= \frac{1}{2}\nu|z_+ z_-|m_B \end{aligned}$$

We find the following relations hold between I_m and m_B in the binary solution, depending on the stoichiometry of the solute formula unit:

- For a 1:1 electrolyte, e.g., NaCl or HCl: $I_m = m_B$
- For a 1:2 or 2:1 electrolyte, e.g., Na_2SO_4 or $CaCl_2$: $I_m = 3m_B$
- For a 2:2 electrolyte, e.g., $MgSO_4$: $I_m = 4m_B$
- For a 1:3 or 3:1 electrolyte, e.g., $AlCl_3$: $I_m = 6m_B$
- For a 3:2 or 2:3 electrolyte, e.g., $Al_2(SO_4)_3$: $I_m = 15m_B$

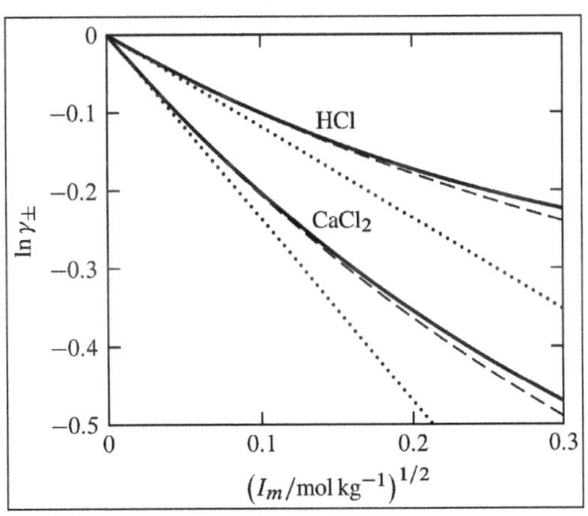

Figure: Dependence of $\ln\gamma\pm$ on $\sqrt{I_m}$ for aqueous HCl (upper curves) and aqueous $CaCl_2$ (lower curves) at 25 °C. Solid curves: experimental; dashed curves: Debye–Hückel equation (a=5×10^{-10}m for HCl, a=4.5×10^{-10}m for $CaCl_2$); dotted lines: Debye–Hückel limiting law.

Figure shows $\ln\gamma\pm$ as a function of $\sqrt{I_m}$ for aqueous HCl and $CaCl_2$. The experimental curves have the limiting slopes predicted by the Debye–Hückel limiting law, but at a low ionic strength the curves begin to deviate significantly from the linear relations predicted by that law. The full Debye–Hückel equation fits the experimental curves over a wider range of ionic strength.

The Gibbs-Helmholtz Equation

The name Gibbs-Helmholtz is usually associated with the equation:

$$H = G - T(\partial G / \partial T)_P$$

w here, H is enthalpy, G is Gibbs free energy and T temperature, with the implied assumption that the differentiation is to be carried out at constant composition. However, this is just one of a very large number of equations of similar form, which, at various times have been called Gibbs-Helmholtz equations. The equation given is, however, most appropriately called the Gibbs-Helmholtz equation since it was first recognized by both Gibbs and Helmholtz (independently).

The greatest utility of this equation lies in the presence of the temperature derivative of the Gibbs free energy. For example, the equilibrium constant in a chemical reaction is determined by the standard Free Energy change for the reaction. Thus, knowledge of the temperature derivative of the free energy provides information relating the temperature derivative of the equilibrium constant. The equilibrium constant, K, may be written as:

$$\Delta g^{\circ} = -\tilde{R} T \ln K$$

where, Δg° is the standard Gibbs free energy change for the reaction at temperature T and $\tilde{R}$ is the universal gas constant.

Applying the Gibbs-Helmholtz equation to each component in the reaction gives:

$$\Delta h^{\circ} = \Delta g^{\circ} - T\left(\partial \Delta g^{\circ} / \partial T\right)_P$$

and hence,

$$(d \ln K)/(d[1/T]) = \Delta h^{\circ} / \tilde{R}$$

Derivation of the Gibbs–Helmholtz Equation

Gibbs–Helmholtz (G-H) equation:

$$\left[\frac{\partial G / T}{\partial T}\right]_P = \frac{-H}{T^2}$$

The Gibbs–Helmholtz equation expresses the temperature dependence of the ratio of G/T at constant pressure, which is a composite function of T as G itself also depends on the temperature.

This fundamental equation is very important, since it is the starting point to the van't Hoff equation, expressing the temperature dependence of the equilibrium constant that interprets quantitatively the shift of chemical equilibrium upon temperature change, predicted in the Le Chatelier–Brown principle.

In a widely used physical chemistry, the starting point to the G–H equation is the definition of Gibbs' free energy, and the relation between its temperature coefficient and entropy:

$$G = H - TS; -S = \left(\frac{\partial G}{\partial T}\right)_P$$

As a next step, these two equations are combined to result:

$$\left(\frac{\partial G}{\partial T}\right)_P = \frac{G-H}{T}$$

and a transformation gives:

$$\left(\frac{\partial G}{\partial T}\right)_P - \frac{G}{T} = -\frac{H}{T}$$

Then it is suggested to recognize the validity of the equation:

$$\left(\frac{\partial G}{\partial T}\right)_P - \frac{G}{T} = T\left(\frac{\partial}{\partial T}\frac{G}{T}\right)_P$$

However, this conclusion is not self-evident for the average chemistry student, as its approval needs not only the derivation of a composite function:

$$\left(\frac{\partial}{\partial T}\frac{G}{T}\right)_P = \frac{1}{T}\left(\frac{\partial G}{\partial T}\right)_P + G\left(\frac{\partial}{\partial T}\frac{1}{T}\right)_P$$

but also a comparison with the rearranged form of $\left(\frac{\partial G}{\partial T}\right)_P - \frac{G}{T} = -\frac{H}{T}$:

$$\left(\frac{\partial}{\partial T}\frac{G}{T}\right)_P = \frac{1}{T}\left[\left(\frac{\partial G}{\partial T}\right)_P - \frac{G}{T}\right] = \frac{-H}{T^2}$$

As it can be seen, the train of thought is not totally straightforward, which hampers the logic of the derivation. According to classroom experiences, the derivation rises difficulties in the sense that it is not so easy to realize that the left side of equation $\left(\frac{\partial G}{\partial T}\right)_P - \frac{G}{T} = T\left(\frac{\partial}{\partial T}\frac{G}{T}\right)_P$ is the product of T with the derivative of G/T.

Some years ago during the Physical Chemistry course, one of the then students of us suggested the following derivation: let us take the definition of G:

$$G = H - TS$$

and divide both sides by T:

$$\frac{G}{T} = \frac{H}{T} - S$$

If we derive this equation with respect to T at constant pressure,

$$\left[\frac{\partial\left(\frac{H}{T}-S\right)}{\partial T}\right]_P=\frac{1}{T}\left(\frac{\partial H}{\partial T}\right)_P-\frac{H}{T^2}-\left(\frac{\partial G}{\partial T}\right)_P$$

We receive equation above consisting of three terms. Comparing it with

$\left[\frac{\partial G/T}{\partial T}\right]_P=\frac{-H}{T^2}$, it can be realized that the first and the third terms are missing, they assumingly have cancelled each other. Seemingly, it might be also the result if H and S were independent of the temperature. Since it is evidently not the case recognized also by himself, this striking coincidence prompted us to find an easily comprehensible proof, and to show the students through a simple reasoning, why the sum of these two terms in $\left[\frac{\partial\left(\frac{H}{T}-S\right)}{\partial T}\right]_P=\frac{1}{T}\left(\frac{\partial H}{\partial T}\right)_P-\frac{H}{T^2}-\left(\frac{\partial G}{\partial T}\right)_P$ gives zero.

For the first term we have:

$$\frac{1}{T}\left(\frac{\partial G}{\partial T}\right)_P=\frac{C_P}{T}$$

In order to receive the third term, we consider that for the infinitesimal change in entropy at constant pressure we have:

$$dS=\frac{dq_{rev}}{T}=\frac{C_P dT}{T}$$

Equation $dS=\frac{dq_{rev}}{T}=\frac{C_P dT}{T}$ implies that:

$$\left(\frac{\partial S}{\partial T}\right)_P=\frac{C_P}{T}$$

So, that we obtained the same expression for the first and third element in:

$$\left[\frac{\partial\left(\frac{H}{T}-S\right)}{\partial T}\right]_P=\frac{1}{T}\left(\frac{\partial H}{\partial T}\right)_P-\frac{H}{T^2}-\left(\frac{\partial G}{\partial T}\right)_P$$

Since the first and third term at the right side in this equation are identical but with opposite sign, they are cancelled.

Thus, last few equations represent a direct, straightforward path to obtain G-H equation from the definition of G in five steps. We are aware that in one of the referred textbooks the derivation through last few equations and previous equation is presented. However, McQuarrie and Simon have to refer also to previous equations to prove validity of $\left(\frac{\partial S}{\partial T}\right)_P = \frac{C_P}{T}$, where they come to the conclusion in four further steps. Thus, the derivation consists altogether of eight elements. We emphasize that our derivation in five steps includes the concrete proof that the sum of the first and third term in:

$$\left[\frac{\partial\left(\frac{H}{T}-S\right)}{\partial T}\right]_P = \frac{1}{T}\left(\frac{\partial H}{\partial T}\right)_P - \frac{H}{T^2} - \left(\frac{\partial G}{\partial T}\right)_P .$$

It is illuminating to point out that—assuming that H and S were temperature independent quantities—we could derive the Gibbs–Helmholtz equation in one step from equation above. But the assumption is evidently wrong. The coincidence, however, reveals that the first derivatives of H and S with respect to temperature are related, and this relation results in the elimination of the first and third terms in equation above.

Nernst Distribution Law

This Law was given by Nernst. This law gives the relationship between the concentrations of a given substance in two different phases in equilibrium with each other.

Suppose, if a small quantity of solute soluble in both the liquids is added then solute distributes itself in two liquid & equilibrium is set up. Both the liquids must be immiscible with each other. If the concentration of the solute in two liquids are C_1 & C_2. According to this law,

$C_1/C_2=K$ (constt.)

where, k= distribution or partition coefficient.

Factors affecting Distribution Coefficient:

- The value of K depends upon the temperature, nature of solute, nature of two solvents .
- It does not depend upon the amount of solute or solvents taken.

If the solubilities of given solute in the two solvents at the given temperature are S_1 & S_2 respectively, then,

$C_1/C_2 = S_1/S_2 = K$

Example: Water & ether are immiscible with each other. Solute succinic acid is soluble in both of these solvents. If small amount of succinic acid is added in a mixture. Then:

Concentration of succinic acid in water = C_{water}

Concentration of succinic acid in ether = C_{ether}

$K = C_{water} / C_{ether}$

C_{water} is more than C_{ether} so C_{water} is taken in the numerator & C_{ether} in the denominator.

Experimental Verification

For experimental verification of distribution law, different quantity of the solute are added in different quantities of two solvents in many experiments. If the value of distribution coefficient for each experiment remains the same. This shows that the data follows distribution law.

Conditions for Distribution Law

- Both the solvents must be immiscible with each other.
- It is applicable only for dilute solution.
- Temperature remains constant throughout the experiment.
- The molecular state of the solute in the two solvents should be the same.

Applications and Limitations of Distribution Law

The distribution law can be applied to a number of physical and chemical processes. Thus the association of the solute in one phase can be ascertained, and degree of association and therefore, the molecular mass of the solute in a given liquid can be found out. The law may also be used to determine the degree of dissociation of acids, bases or salts in a solvent. Other examples are given below:

a. The degree of hydrolysis of substances can be conveniently studied. The hydrolysis of aniline hydrochloride in water can be followed from the distribution of aniline hydrochloride between water and benzene. Thus for this reaction:

Aniline hydrochloride (Salt) + Water (H_2O) ↔ Aniline (Base) + Hydrochloric acid (Acid)

Thus,

$K_h = \{[Base] \times [Acid]\} / \{[Salt] \times [H_2O]\}$

By applying the principle of equilibrium it can be shown that,

$K_h = \{C_1^2 (1+vK)\} / \{C_2 - C_1(1+vK)\}$

where, C_1 is the concentration of the base in aqueous layer (C_1 and C_2 are expressed in Mol L^{-1}), C_2 the original concentration of the salt, K the distribution coefficient and v is the volume of benzene in liters added to 1 liter of water.

b.The formula of the complex $CuSO_4.4NH_3$ was confirmed from the studies of the distribution of ammonia between chloroform and aqueous Copper Sulphate Solution.

c. Another example of the application of the distribution law can be demonstrated by finding the equilibrium constant for the equilibrium.

Determination of Solubility

The value of K is equal to the ratio of solubility of the solute in the two solvents.

$$K = C_1/C_2 = S_1/S_2$$

If the value of K and solubility of the solute in one solvent is known then we can calculate the solubility of a solute in other solvents.

Solvent Extraction

Extraction of one substance from a solution containing various substances by using a suitable solvent is known as solvent extraction. "The amount of extracted substance is more if a smaller amount of solvent is used many times rather than using a larger amount of solvent a fewer times".

Real-life Applications of Distribution Law

- Solvent removal;
- Separation chromatography;
- A release of drug from quantity forms;
- Conservation of emulsions and creams;
- Formation of a solubilized structure;
- One substance can be extracted from the solution containing a variety of substances;
- Used in high act liquid chromatography;
- The solubility of drugs in water and other solvents and in a mixture of solvents can be predicted;
- Structure-activity relationship (SAR) for a series of drugs can be studied.

Limitations of Distribution Law

- Stable temperature: The temperature is kept stable during the experiment.
- Same molecular state: The law does not hold if there is an association or dissociation of the solute in one of the solvents. The molecular state of the solute has to stay stable when in make contact with the solvent. It should not endure dissociation or involvement.
- Equilibrium concentrations: The concentrations of the solute are noted after the balance has been established.
- Dilute solutions: The concentration of the solute in the two solvents is low. The law does not hold when the concentrations are high. The solute that is being dispersed shall not on ant situation imprudent towards the solvents being used.
- Among the applications of Nernst's distribution law, the most significant and attractive one is the extraction of a material dissolved in one solvent by another solvent, the two solvents being partially miscible or immiscible.

Phase Rule

Gibbs Phase Rule is expressed by the simple formulation:

$$P + F = C + 2,$$

where,

P is the number of phases in the system.

A phase is any physically separable material in the system. Every unique mineral is a phase (including polymorphs); igneous melts, liquids (aqueous solutions), and vapor are also considered unique phases. It is possible to have two or more phases in the same state of matter (e.g. solid mineral assemblages, immiscible silicate and sulfide melts, immiscible liquids such as water and hydrocarbons, etc.) Phases may either be pure compounds or mixtures such as solid or aqueous solutions--but they must "behave" as a coherent substance with fixed chemical and physical properties.

C is the minimum number of chemical components required to constitute all the phases in the system.

For historical reasons, geologists normally define components in terms of the simple oxides (e.g. SiO_2, Al_2O_3, CaO, etc.). If two possible components always occur in the same proportions in multiple phases in a system, these can be combined into a single component (remember, we are always trying to define the *minimum* number of components required to make all the phases in the system). Consider the reaction:

$$CaMg(CO_3)_2 + 2\ SiO_2 = CaMgSi_2O_6 + 2\ CO_2$$

Dolomite + 2 Quartz = Diopside + 2 Carbon Dioxide

Normally we would pick the 4 components: CaO, MgO, SiO_2, and CO_2 based on the simple oxides. However, because Ca and Mg are in a 1:1 ratio in both dolomite and diopside (and not present in quartz or carbon dioxide), we can consider this a ternary system with components: $CaMgO_2$, SiO_2, and CO_2.

In some geologic systems it is convenient to define the components in terms of end-member compositions (e.g. binary systems such as carbonates, $CaCO_3$ - $MgCO_3$; alkali feldspars, $NaAlSi_3O_8$ - $KAlSi_3O_8$, etc.).

In some cases, a given mineral assemblage may be represented by a sub-set of the whole system if fewer components are needed to define the compositions of the observed mineral assemblage--this is known as a *degenerate* system. A good example of this can be seen on a ternary (3 component) chemographic projection if the phases of interest plot in a co-linear fashion (i.e. requiring only two components to define the phase compositions in what is otherwise a three component system).

F is the number of degrees of freedom in the system (also referred to as the variance of the system). For geologic applications, this generally refers to the number of variables (e.g. pressure and temperature) that can be independently changed without altering the state of the system (i.e. the number of phases and their compositions are constant). Three common types of equilibria are possible:

- Invariant equilibria, in which neither P nor T can be changed; on a phase diagram, this is represented as a singular *invariant point*.
- Univariant equilibria, in which either P or T can be changed independently, but to maintain the state of the system, there must be a corresponding change in the other variable; on a phase diagram this is referred to as a *univariant curve*.
- Divariant equilibria, in which both P and T are free to change independently without changing the state of the system (but bounded by the conditions defined by the univariant equilibria).

The Integer in the Gibbs Phase Rule

The integer in the Gibbs Phase Rule is related to the number of *intensive* parameters (i.e. those that are independent of mass; such as pressure and temperature) that are being considered. Note that many phase diagrams consider only one intensive parameter (e.g. T-X_{fluid}, where T is intensive, and the mole fraction of fluid is related to the amount of mass of the fluid components); in this case, the Gibbs Phase Rule would be: P + F = C + 1.

It is generally the case that as the number of components increase, the variance of the system must also increase. Conversely, as the number of phases increase in the system, the variance of the system must correspondingly decrease. A corollary to Gibbs' Phase Rule is Goldschmidt's Mineralogical Phase Rule: for a given rock in equilibrium at a fixed P and T, the number of phases is less than or equal to the number of components. Geologic systems are really rather simple! This means that we have a reasonable expectation that we can successfully interpret the conditions of formation of rocks in their natural settings, and to model their behavior using the thermodynamic databases and programs presented at the end of this module.

Component Examples

Let's consider the simple 1 component system for H_2O:

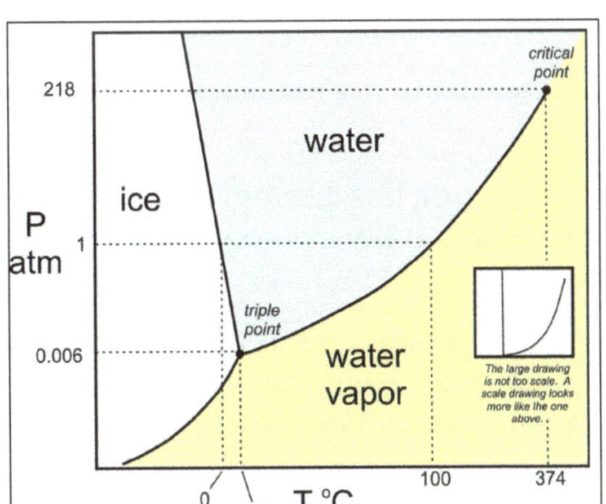

Phase diagram for the one component system H_2O. This diagram is not to scale; a scale version looks like the thumbnail in the white box.

- The system is entirely composed of H_2O, so there is only one component present.
- The phases present represent three states of matter: liquid (water), solid (ice), and vapor (steam). All have distinct physical properties (e.g. density, structure--or lack of, etc.) and chemical properties (e.g. $\Delta G_{formation,}$ molar volume etc.) so they must be considered distinct phases.
- Note that there is only one point on this diagram where all three phases coexist in equilibrium--this "triple point" is also referred to as an *invariant point*; because P and T are uniquely specified, there are zero degrees of freedom.
- Each of the curves represents a chemical reaction that describes a phase transformation: solid to liquid (melt/crystallization), liquid to vapor (boiling/condensation), solid to vapor (sublimation/deposition). There are three *univariant curves* around the invariant point; it is always the case that for a C-component system, there will always be C+2 univariant curves radiating around an invariant point. There is only one degree of freedom along each of the univariant curves: you can independently change either T or P, but to maintain two coexisting phases along the curve the second variable must change by a corresponding fixed amount.
- There are three distinct areas where only ice, liquid, or vapor exit. These are *divariant* fields. T and P are both free to change within these fields and you will still have only one phase (a bit hotter or colder, or compressed or expanded, but nonetheless the same phase).
- The end of the "boiling curve", separating the liquid to vapor transition, is called the "critical point". This is a particularly interesting part of the phase diagram because beyond this region the physico-chemical properties of water and steam converge to the point where they are identical. Thus, beyond the critical point, we refer to this single phase as a "supercritical fluid".

Now, let's consider a simple 1 component system that describes the mineral phases in the aluminosilicate system:

- The entire system is defined by one component: Al_2SiO_5 (i.e. all the phases can be completely made of this one component).
- There are three solid phases shown in this diagram: the polymorphs of Al_2SiO_5 andalusite, kyanite and sillimanite.
- There is only one unique place on this diagram where all three phases can coexist in equilibrium-the invariant point at 3.8 Kb and 500 °C; at this point there are zero degrees of freedom.
- There are three univariant reactions on this diagram, each representing the phase transitions: andalusite = sillimanite, andalusite = kyanite, and kyanite = sillimanite. In each of these reactions, *either* pressure or temperature can be changed independently, but for the state of the system to remain the same (i.e. two solid phases coexisting in equilibrium), the other variable must change by a fixed amount to maintain the assemblage on the univariant curve-so there is one degree of freedom.

- There are three divariant fields in which only a single mineral phase is stable. Within these fields pressure *and* temperature may be changed independently without changing the state of the system-thus there are two degrees of freedom in the divariant fields.

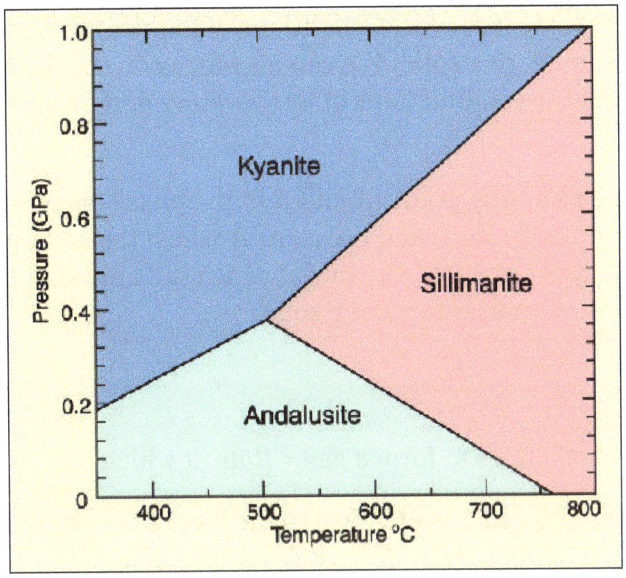

Phase diagram for the one component system Al_2SiO_5.

The figure below shows an example of a phase diagram, which summarizes the effect of temperature and pressure on a substance in a closed container. Every point in this diagram represents a possible combination of temperature and pressure for the system. The diagram is divided into three areas, which represent the solid, liquid, and gaseous states of the substance.

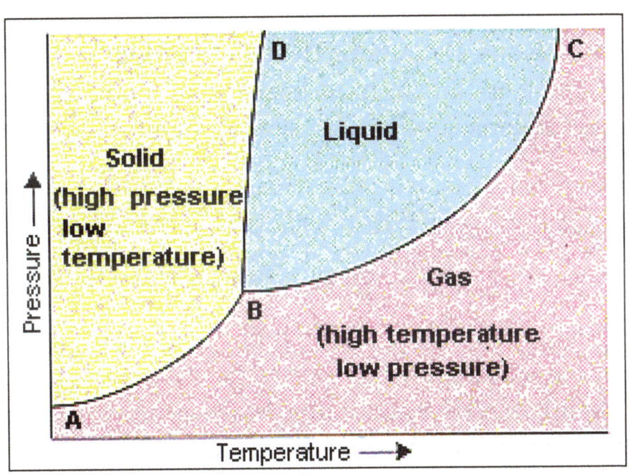

The best way to remember which area corresponds to each of these states is to remember the conditions of temperature and pressure that are most likely to be associated with a solid, a liquid, and a gas. Low temperatures and high pressures favor the formation of a solid. Gases, on the other hand, are most likely to be found at high temperatures and low pressures. Liquids lie between these extremes.

You can therefore test whether you have correctly labeled a phase diagram by drawing a line from left to right across the top of the diagram, which corresponds to an increase in the temperature of the system at constant pressure. When a solid is heated at constant pressure, it melts to form a liquid, which eventually boils to form a gas.

Phase diagrams can be used in several ways. We can focus on the regions separated by the lines in these diagrams, and get some idea of the conditions of temperature and pressure that are most likely to produce a gas, a liquid, or a solid. We can also focus on the lines that divide the diagram into states, which represent the combinations of temperature and pressure at which two states are in equilibrium.

The points along the line connecting points *A* and *B* in the phase diagram in the figure above represent all combinations of temperature and pressure at which the solid is in equilibrium with the gas. At these temperatures and pressures, the rate at which the solid sublimes to form a gas is equal to the rate at which the gas condenses to form a solid.

Along AB line

Rate at which solid sublimes to form a gas = Rate at which gas condenses to form a solid.

The solid line between points *B* and *C* is identical to the plot of temperature dependence of the vapor pressure of the liquid. It contains all of the combinations of temperature and pressure at which the liquid boils. At every point along this line, the liquid boils to form a gas and the gas condenses to form a liquid at the same rate.

Along BC line

Rate at which liquid boils to form a gas = Rate at which gas condenses to form a liquid.

The solid line between points *B* and *D* contains the combinations of temperature and pressure at which the solid and liquid are in equilibrium. At every point along this line, the solid melts at the same rate at which the liquid freezes.

Along BD line

Rate at which solid melts to form a liquid = Rate at which liquid freezes to form a solid.

The *BD* line is almost vertical because the melting point of a solid is not very sensitive to changes in pressure. For most compounds, this line has a small positive slope, as shown in the figure above. The slope of this line is slightly negative for water, however. As a result, water can melt at temperatures near its freezing point when subjected to pressure. The ease with which ice skaters glide across a frozen pond can be explained by the fact that the pressure exerted by their skates melts a small portion of the ice that lies beneath the blades.

Point *B* in this phase diagram represents the only combination of temperature and pressure at which a pure substance can exist simultaneously as a solid, a liquid, and a gas. It is therefore called the triple point of the substance, and it represents the only point in the phase diagram in which all three states are in equilibrium. Point *C* is the critical point of the substance, which is the highest temperature and pressure at which a gas and a liquid can coexist at equilibrium.

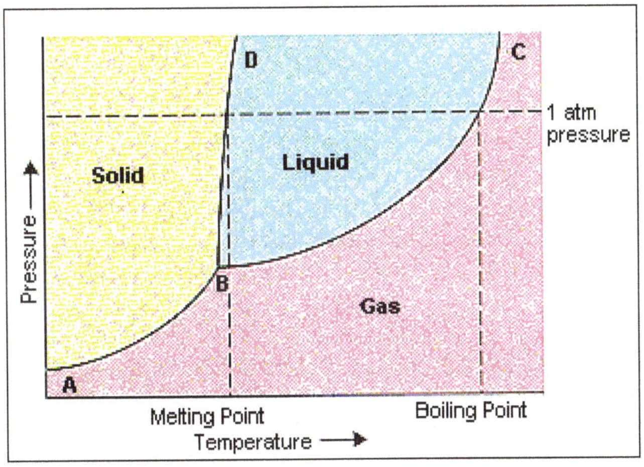

The figure shows what happens when we draw a horizontal line across a phase diagram at a pressure of exactly 1 atm. This line crosses the line between points *B* and *D* at the melting point of the substance because solids normally melt at the temperature at which the solid and liquid are in equilibrium at 1 atm pressure. The line crosses the line between points *B* and *C* at the boiling point of the substance because the normal boiling point of a liquid is the temperature at which the liquid and gas are in equilibrium at 1 atm pressure and the vapor pressure of the liquid is therefore equal to 1 atm.

References

- Principles-of-Chemical-Equilibrium, Principles-of-Chemical-Equilibria, Chemical-Equilibria, Equilibria, Physical-and-Theoretical-Chemistry: libretexts.org, Retrieved 5 March, 2019
- The-equilibrium-constant, introductorychemistry: opentextbc.ca, Retrieved 7 May, 2019
- Definition-of-le-chateliers-principle: thoughtco.com, Retrieved 17 March, 2019
- The-Debye-HBCckel-Theory, Electrolyte-Solutions, DeVoe's-Thermodynamics-and-Chemistry, Physical-and-Theoretical-Chemistry: libretexts.org, Retrieved 13 January, 2019
- On-the-derivation-of-the-Gibbs-Helmholtz-equation: researchgate.net, Retrieved 3 August, 2019
- Distribution-law: chemistryonline.guru, Retrieved 22 June, 2019
- Applications-distribution-law, chemistry: qsstudy.com, Retrieved 11 February, 2019
- Phaserule, equilibria: carleton.edu, Retrieved 10 April, 2019

Permissions

All chapters in this book are published with permission under the Creative Commons Attribution Share Alike License or equivalent. Every chapter published in this book has been scrutinized by our experts. Their significance has been extensively debated. The topics covered herein carry significant information for a comprehensive understanding. They may even be implemented as practical applications or may be referred to as a beginning point for further studies.

We would like to thank the editorial team for lending their expertise to make the book truly unique. They have played a crucial role in the development of this book. Without their invaluable contributions this book wouldn't have been possible. They have made vital efforts to compile up to date information on the varied aspects of this subject to make this book a valuable addition to the collection of many professionals and students.

This book was conceptualized with the vision of imparting up-to-date and integrated information in this field. To ensure the same, a matchless editorial board was set up. Every individual on the board went through rigorous rounds of assessment to prove their worth. After which they invested a large part of their time researching and compiling the most relevant data for our readers.

The editorial board has been involved in producing this book since its inception. They have spent rigorous hours researching and exploring the diverse topics which have resulted in the successful publishing of this book. They have passed on their knowledge of decades through this book. To expedite this challenging task, the publisher supported the team at every step. A small team of assistant editors was also appointed to further simplify the editing procedure and attain best results for the readers.

Apart from the editorial board, the designing team has also invested a significant amount of their time in understanding the subject and creating the most relevant covers. They scrutinized every image to scout for the most suitable representation of the subject and create an appropriate cover for the book.

The publishing team has been an ardent support to the editorial, designing and production team. Their endless efforts to recruit the best for this project, has resulted in the accomplishment of this book. They are a veteran in the field of academics and their pool of knowledge is as vast as their experience in printing. Their expertise and guidance has proved useful at every step. Their uncompromising quality standards have made this book an exceptional effort. Their encouragement from time to time has been an inspiration for everyone.

The publisher and the editorial board hope that this book will prove to be a valuable piece of knowledge for students, practitioners and scholars across the globe.

Index

Q

R

S

T

V

W

CPSIA information can be obtained
at www.ICGtesting.com
Printed in the USA
LVHW060856030122
707669LV00002B/191